SCIENTISTS CONFRONT CREATIONISM

Edited by Laurie R. Godfrey

SCIENTISTS CONFRONT CREATIONISM

W. W. NORTON & COMPANY
New York · London

Published simultaneously in Canada by George J. McLeod Limited, Toronto.
Printed in the United States of America.

The text of this book is composed in 10/12 Avanta, with display type set in Horizon Light.
Manufacturing by the Haddon Craftsmen, Inc.
Book design by Nancy Dale Muldoon.

First Edition

Library of Congress Cataloging in Publication Data
Main entry under title:

Scientists confront creationism.

Includes bibliographical references and index.
1. Creationism—Controversial literature—
Addresses, essays, lectures. 2. Evolution—
Addresses, essays, lectures. I. Godfrey, Laurie R.
BS652.G34 1983 231.7'65 82-12500

ISBN 0-393-01629-3

W. W. Norton & Company, Inc., 500 Fifth Avenue, New York, N. Y. 10110
W. W. Norton & Company Ltd., 37 Great Russell Street, London WCIB 3NU

1 2 3 4 5 6 7 8 9 0

To my family—Paul, Darren, and Mollie;
and to parents, like Marian Finger,
who care about their children's
science education.

Contents

Preface

The Arkansas law requiring "balanced treatment for creation science and evolution science" was declared unconstitutional on 5 January 1982. Even so, public support for "equal time" is fermenting in Arkansas and in many other states of the Union. Without legal mandates, some school districts are nevertheless using creationist films, cassettes, and texts purporting to present only "scientific evidence" for creation in their science curriculums. Simultaneously, in school districts seemingly not affected by creationist "two-model" politics, the subject of evolution is quietly disappearing from textbooks and curriculums. The body of "scientific" misinformation that constitutes the core of "creation science" is reaching a wider cross-section of the American populace and threatening the scientific literacy of its youth.

Anyone familiar with the history of American antievolutionism and the arguments of the nineteenth-century fundamentalist churches against "science falsely so-called" will recognize in modern scientific creationism an old script played out on a new stage. The rhetoric may depart from that best remembered of the Scopes era: William Jennings Bryan's derision of concern for the "age of rocks" instead of the rock of ages, or his famous indictment of those who would begin with a piece of dirt and reason up, rather than begin with God and reason down. Nevertheless, even the militant "pro-science" rhetoric spouted by the leaders of the modern scientific creationist movement is rooted in the rhetoric of some Scopes era crusaders—people like George McCready Price and Byron Nelson. Indeed, the roots of modern "scientific creationism" lie further back in time—in the literature of nineteenth-century fundamentalist churches. Seventh-Day Adventists Ellen White, Uriah Smith, and Alonzo T. Jones professed no disdain for "real" science; they preferred, rather, to redefine it. For they averred that the "facts" of science could not possibly contradict the Bible, interpreted as the literal, infallible, and perspicuous Word of God. Ellen White proclaimed that "the Bible is not to be tested by men's ideas of science, but science is to be brought to the test of this unerring standard" (Numbers 1975). More than half a century later, Adventist George McCready Price (1941) would remain true to his religious tradition in characterizing as anti-Christian those who pretended to have knowledge contrary to the revealed will of God. If "true" science—the "facts" of science—could not contradict the Bible, then apparent contradictions must emanate not from *science,* but from *scientists* who, forgetting the standard against which their ideas must be judged, would play a role in creating the final delusions that would take the world captive, as portrayed in Revelations. Controversy within this "false" science betrays its inherent weakness, for "theories" are the mere fallible "ideas" of men.

It was thus in the late nineteenth century and early twentieth century that the tenets of geology became, to the minds of fundamentalists, mere "guesswork," lacking scientific "proof." Darwin became a hopeless and frivolous blunderer whose lack of faith guided his scientific errors. In the words of George McCready Price, Charles Darwin possessed a "slow lumbering type [of mind] so characteristic of the average hunting and sporting country squire of England" (1930, p. 88); after all "Darwin was never more than a prattling schoolboy" (Price 1930, p. 89). Darwin did not understand the "facts" of science.

Modern scientific creationists treat Charles Darwin with the same deference; they admit knowing no "fact"—not even one—that contradicts their literal reading of Genesis. This they will defend even if it requires rejecting the most basic tenets of all sciences and replacing them with miracles and an entirely new set of "facts." Like their forebears of the nineteenth century, they would rely upon a "universal flood" to resolve apparent conflicts between science and Genesis. Such a postulate, they say, reveals simultaneously the fallacy of the concept of evolution and of other fundamentalist biblical apologetics (such as the notion that a "day" in Genesis represents a long period of time, as required by the fossil evidence, or the notion that fossils are remnants of a world that existed before God created the present order in six days). Modern "scientific creationists" are indeed resurrecting ghosts from the nineteenth century. Many of the same old arguments against evolution and an old earth, refuted time and time again, are being revived by Christian fundamentalists.

Within decades of the publication of Charles Darwin's *The Origin of Species,* many theologians had embraced evolution. In today's world —with all we know about the sources, translations, and changing historical interpretations of the Bible, with all we know about the function of religious myths in different cultural contexts, and with all we know about the facts of science—it would seem patently ridiculous to maintain that the Bible is a historically and scientifically accurate account of actual physical events. But scientific creationists not only insist upon this interpretation, they are attempting to force its public indoctrination. They represent evolutionists as defenders of a crumbling orthodoxy who are secretly aware that evolution is a dead horse, but who are unable, because of their "religion," to admit it. In a recent radio interview sponsored by the Institute for Creation Research, creationist Richard Bliss declared:

> The anti-creationists (and these are distinct from those evolutionary scientists and persons who are open-minded and willing to be convinced of the facts) have declared war against reason, both logical and scientific. . . . They are saying that science doesn't dare to pursue truth wherever it lies. . . . Random processes, without God, are the faith and substance of science, even if the facts of science point in another direction. . . . The anti-creationist is desperate. He cannot win the scientific argument and is not willing to concede its failure. I cannot blame those who have spent their lifetime with atheism and evolution for being disturbed as they watch their castles fall apart, but the anti-creationists have a deep and bitter vendetta against the obvious truths of science. . . .

But it is the "scientific creationists," not evolutionists, who are defending an old orthodoxy.

Scientific creationism is not science; it is religion. First, the "creation model" mirrors Genesis and no other creation myth. Second, creation science uses biblical terms, such as "kind," that have no meaning in science. Third, creationists invoke biblical miracles to "explain" all phenomena, while making no pretense of understanding God's power or methods. Raindrop impressions in rock become signs, not of a rainstorm, but of a global hydraulic cataclysm. Dinosaur footprints reveal giant reptiles striving to escape the rampages of the Flood. Sets of giant humanlike footprints (known to have been carved by townsfolk during the Depression) demonstrate the contemporaneity of man and dinosaurs. Indeed, their giantism becomes testimony to the luxurious health of antediluvian man. Protected by the "canopy" of water vapor or ice crystals from the harmful radiation of the sun, these giants must have lived long lives and had large families. They did so, that is, until God unleashed waters from the canopy above and the springs of the deep—a miracle that "explains" the burial, fossilization, and indeed extinction of many of God's less worthy created "kinds."

The teaching of "scientific creationism" as science in the public schools of America represents a flagrant violation of the principle of separation of church and state. Far from opening educational horizons to competing alternate explanations, scientific creationists construct artificial dichotomies that distort and confound the methods and functions of science and religion. They ally all religions that depart from their own biblical literalism with "evolution." They pretend that biblical literalism is the logical alternative to evolution. They mold the "facts" of science according to their needs. They distort the meanings of "fact" and "theory." Cloaked in scientific jargon but uninformed by the scientific developments of the past one hundred fifty years, the arguments of the scientific creationists demand a response. Such is the task of *Scientists Confront Creationism.*

Laurie R. Godfrey, *Editor*

REFERENCES CITED

Numbers, Ronald L. 1975. Science falsely so-called: evolution and Adventists in the nineteenth century. *Journal of the American Scientific Affiliation,* 27: 18–23.

Price, George McCready. 1930. *A history of some scientific blunders.* New York: Fleming H. Revell Co.

—. 1941. *Genesis vindicated.* Takoma Park, Washington, D. C.: Review and Herald Publ. Assoc.

Contributors

George O. Abell is professor of astronomy at the University of California, Los Angeles. He received his Ph. D. from the California Institute of Technology and has served as guest investigator at the Hale Observatories, visiting scientist at the Max-Planck Institute of Physics and Astrophysics, and as visiting professor at the Royal Observatory in Edinburgh. Dr. Abell has published numerous professional papers and articles on astronomy and is the author of several books on the subject, including *Exploration of the Universe* (4th edition, Saunders College Publishers, 1982), *Drama of the Universe* (Holt, Rinehart and Winston, 1978), and *Realm of the Universe* (Saunders College Publishers, 1980). He is the editor of *Science and the Paranormal* (Scribner's, 1981, with Barry Singer) and is a member of the executive council of the Committee for the Scientific Investigation of Claims of the Paranormal.

C. Loring Brace is professor of anthropology and curator of physical anthropology at the Museum of Anthropology, University of Michigan, Ann Arbor. Dr. Brace has published extensively on human evolution and is the author of several books on the subject, including *The Stages of Human Evolution* (Prentice-Hall, 1969), *An Atlas of Human Evolution* (Holt, Rinehart and Winston, 1979, with Harry Nelson, Noel Korn, and Mary L. Brace), and *Human Evolution: An Introduction to Biological*

Anthropology (Macmillan, 1977, with Ashley Montagu). He is especially well known for his arguments favoring the inclusion of Neanderthals in the mainstream of human evolution. Dr. Brace is also interested in the intellectual history of biological anthropology and its influence on modern interpretations of human evolution.

Stephen G. Brush received his D. Phil. in theoretical physics from Oxford University. He is currently professor of history of science in the Department of History and Institute for Physical Science and Technology at the University of Maryland, College Park. He has also worked as a physicist at the Lawrence Radiation Laboratory and has taught at Harvard University. He has published extensively in physics and history of science and is the author of *Kinetic Theory* (three volumes, Pergamon, 1965), *The Kind of Motion We Call Heat: A History of the Kinetic Theory of Gases in the Nineteenth Century* (Elsevier, 1976), and *The Temperature of History: Phases of Science and Culture in the Nineteenth Century* (B. Franklin, 1978). Dr. Brush is widely known for his research into the history of physical science in the nineteenth and twentieth centuries.

John R. Cole received his Ph. D. in anthropology from Columbia University and now teaches anthropology in the Department of Sociology and Anthropology at the University of Northern Iowa in Cedar Falls. He is a consulting editor of *The Skeptical Inquirer* magazine and co-chairman of the education subcommittee of the Committee for the Scientific Investigation of Claims of the Paranormal. An archaeologist with field experience on four continents, Dr. Cole's main interests and publications are in New World archaeology, cultural anthropology, and popular science.

Joel L. Cracraft received his Ph. D. in biology from Columbia University. He has been a research fellow at the American Museum of Natural History and has taught since 1970 at the University of Illinois Medical Center where he currently holds the title of associate professor of anatomy. Dr. Cracraft is an expert on the functional morphology of birds, systematic theory, and vertebrate biogeography. Besides numerous professional papers, he is the co-author (with Niles Eldredge) of *Phylogenetic Patterns and the Evolutionary Process* (Columbia University

Press, 1980), and co-editor (with Niles Eldredge) of *Phylogenetic Analysis and Paleontology* (Columbia University Press, 1979).

Russell F. Doolittle received his Ph. D. in biochemistry at Harvard University. He taught at Amherst College in Massachusetts, was a USPHS postdoctoral fellow in Sweden for two years, and now teaches in the Department of Chemistry at the University of California in San Diego, where he holds the position of professor of biochemistry. A former Research Career Development Awardee of the National Heart Institute, Dr. Doolittle's main interests are protein chemistry, molecular evolution, and the origin of life, and he has published extensively in these fields.

Frederick Edwords is Administrator of the American Humanist Association, on the board of the New York Council for Evolution Education, and editor of *Creation/Evolution,* a quarterly devoted entirely to responding to claims of the "scientific creationists." He has participated in debates with creationists and lectured extensively on the topic. He has served as California District Director of the American Humanist Association and as president of the Humanist Association of San Diego. Mr. Edwords is managing editor of the publication *Free Mind,* and he writes a column on the creation/evolution controversy for *The Humanist* magazine.

Laurie R. Godfrey received her Ph. D. in anthropology from Harvard University and now teaches physical anthropology in the Department of Anthropology at the University of Massachusetts in Amherst. She has also taught at Cornell University in Ithaca, New York. Her main areas of interest are primatology, functional anatomy, evolutionary biology, and science education. She has published articles on these topics and, most recently, on the current antievolutionist movement. Dr. Godfrey is a scientific consultant to the Committee for the Scientific Investigation of Claims of the Paranormal and a member of the National Advisory Board of *The Voice of Reason.* She is editing a book on current issues in evolution (Allyn & Bacon).

Stephen Jay Gould is professor of geology at Harvard University and teaches courses in biology, geology, and the history of science. He writes a regular column, "This View of Life," for *Natural History* magazine. Two collections of articles from this column, *Ever Since Darwin: Reflections in Natural History* and *The Panda's Thumb* have been published by W. W. Norton. Dr. Gould is also the author of *Ontogeny and Phylogeny* (Belknap, Harvard, 1977), a book that examines the relationship between individual development and evolutionary change. His latest book *The Mismeasure of Man* (Norton, 1981) analyses misuses of metrics in the published work of "racial historians" and other scientists concerned with variation in human "intelligence." Gould is both a leading popularizer of evolutionary biology and a contributor to the development of the theory of "punctuated equilibria"—an alternative to traditional Darwinian gradualism.

Thomas H. Jukes received his Ph. D. in biochemistry at the University of Toronto. He has held numerous positions in both industry and academia and is currently professor in residence of biophysics and medical physics at the University of California, Berkeley, where he also holds the position of research biochemist at the Space Sciences Lab. Among his numerous awards is an Honorary Doctor of Science degree from the University of Guelph (Ontario). Dr. Jukes is the author of *Antibiotics in Nutrition* (Medical Encyclopedia, Inc., 1955), *Molecules and Evolution* (Columbia University Press, 1966), co-author of *Effects of DDT on Man and Other Mammals* (Irvington, 1973), and author of many articles within his field. He is an expert on nutrition, the genetic code, and exobiology and is one of the most influential scientists in the field of molecular evolution.

Alice B. Kehoe is professor of anthropology at Marquette University in Milwaukee, Wisconsin. She received her Ph. D. in anthropology from Harvard University and has served on the Executive Board of the American Anthropological Association. Her main interests are the ethnology and archaeology of native American peoples. Her most recent book is entitled *North American Indians: A Comprehensive Account* (Prentice-Hall, 1981).

John W. Patterson received his M. S. in mining engineering and Ph. D. in metallurgical engineering from Ohio State University and currently holds the position of professor of engineering in the Department of Materials Science and Engineering at Iowa State University, Ames. His main interests are thermodynamics and solid electrochemistry, and it is in these two fields that he has published most. Dr. Patterson is a scientific consultant to the Committee for the Scientific Investigation of Claims of the Paranormal and a member of its education subcommittee.

David M. Raup is chairman of the department of geophysical sciences at the University of Chicago. He received his Ph. D. in geology from Harvard University; he has taught at Johns Hopkins University, the University of Rochester, and as guest professor at the University of Tubingen in West Germany. He is the former dean of science at the Field Museum of natural history and has served as president of the Paleontological Society. He is also a member of the National Academy of Sciences. Well-known for his research in geology and invertebrate paleontology, Dr. Raup has published extensively in these fields, including the textbook *Principles of Paleontology* (W. H. Freeman, 1978), co-authored with Steven M. Stanley.

Robert J. Schadewald is a freelance science writer from Rogers, Minnesota, who has published numerous essays on pseudoscience and on the history of technology. His essays have appeared in such magazines as *Smithsonian, Science '80, Science Digest, Technology Illustrated, Creation/Evolution,* and *The Skeptical Inquirer.* Mr. Schadewald's main interest is pseudoscience; his special area of expertise is the flat-earth movement (its modern base as well as its cultural and historical roots). Mr. Schadewald has accumulated one of the world's most complete flat-earth libraries.

Steven D. Schafersman received his Ph. D. in geology from Rice University in Houston, Texas, and has taught at several colleges, including the University of Houston. He is currently employed as a research geologist with a major company in the petroleum industry. His main interests and publications are in evolutionary paleontology, carbonate rocks, reefs, and biostratigraphy. Dr. Schafersman is also interested in pseudoscience and the creationist movement and is a member of the education subcommittee of the Committee for the Scientific Investigation of Claims of the Paranormal.

Introduction

The recent massive attack by fundamentalist Christians on the teaching of evolution in the schools has left scientists indignant and somewhat bewildered. Creationist arguments have seemed to them a compound of ignorance and malevolence, and, indeed, there has been both confusion and dishonesty in the creationist attack. First, there has been a confusion, partly deliberate, of the *fact* that organisms have evolved with *theories* about the detailed mechanics of the process. The facts of evolution are clear and are not disputed by any serious scientific worker. The universe is over 11 billion years old, and the earth, in particular, is over 4 billion. That is a fact. Life on earth is at least 2.5 billion years old, and, as new evidence accumulates, the best estimate of the origin of life gets pushed further and further into the past. That too is a fact. It is also a fact that there were no mammals or birds 200 million years ago and no vertebrates 600 million years ago, while there are no dinosaurs now. Finally, it is a fact that under conditions that have existed on earth for the last billion years, at least, all living organisms arise from previously living organisms. So, the present complex living forms have evolved by an unbroken and continuous process from the simplest living forms of the pre-Cambrian era. To assert, on the contrary, that the earth and life on it are a paltry ten or hundred thousand years old and that the complex forms living today arose in an instant from unorganized matter is in contradiction not simply with the corpus of *biological* knowledge but with *all* scientific knowledge of the physical world. To deny evolution is to deny physics, chemistry, and astronomy, as well as biology.

Unlike the facts of evolution, theories about the details of the process are in constant flux, and there is a great deal of disagreement about them. How important is natural selection as opposed to random and historical factors in establishing new features in species? Is all natural selection at the level of individual organisms, or is there also selection among populations and species? Is virtually all mutation on which evolution is based of small magnitude, or have "macro-mutations" played an important part in evolution? Is species formation driven by adaptation, or is it an essentially nonadaptive consequence of sexual isolation between populations? Has evolution proceeded at a fairly regular pace, or are there long periods of stasis, followed by a few hundred or thousand generations of more rapid change that are associated with formation of new species? All of these questions have been and are still hotly debated by students of evolution. They constitute the problems of the *theory* of evolution rather than the *fact* of evolution.

Partly through honest confusion, but also partly through a conscious attempt to confuse others, creationists have muddled the disputes about evolutionary theory with the accepted fact of evolution to claim that even scientists call evolution into question. By melding our knowledge of what has happened in evolution with our doubts about how this has happened into a single "theory of evolution," creationists hope to challenge evolution with evolutionists' own words. Sometimes creationists plunge more deeply into dishonesty by taking statements of evolutionists out of context to make them say the opposite of what was intended. For example, when, in an article on adaptation, I described the outmoded nineteenth-century belief that the perfection of creation was the best evidence of a creator, this description was taken into creationist literature as evidence of my own rejection of evolution. Such deliberate misuse of the literature of evolutionary biology, and the transparent subterfuge of passing off the Old Testament myth of creation as if it were creation "science" rather than the belief of a particular religion, has convinced most evolutionists that creationism is nothing but an ill-willed attempt to suppress truth in the interest of propping up a failing institution. But such a view badly oversimplifies the situation and misses the deep social and political roots of creationism.

Creationism is an American institution, and it is not only American but specifically southern and southwestern. To understand its origin and power, we must understand the situation in which most rural white southerners found themselves in the early years of this century. They were, by and large, tenants or small landholders, barely eking out a living on land owned by others or heavily mortgaged to the banks. They were constantly in debt and had the overwhelming, and quite correct, perception that their lives were under the control of the rich. Whether farmers or miners, they saw themselves as living in a world controlled by the rich northern and eastern bankers and entrepreneurs. Their response to this perception of powerlessness and subjection was twofold. Large numbers became Populists and Socialists. Eugene Debs got more votes in 1912 from the rural, high tenancy counties of Arkansas, Texas, and Oklahoma than from the urban centers where industrial workers were concentrated. The largest selling weekly magazine in the United States in 1910 was the socialist *Appeal to Reason,* published in Gerard, Kansas. At the same time, there was an accent on a rural, revivalist, fundamentalist religion that rejected the sophistries of the rich men's churches. If tenant farmers and small landholders had no control over their land and livelihood, at least they could control their own beliefs and the education of their children.

The exclusion of upper-class eastern and northern culture, the culture of the dominators of society, was successfully maintained until the 1950s. Evolution, for example, was barely mentioned in school textbooks; as late as 1954, my children, in Raleigh, North Carolina, read that "God made the flowers out of sunshine." Then came Sputnik and the demand that the science curriculums be radically revised to make our children scientifically competent. One consequence was the complete rewriting of the biology curriculums by the Biological Science Curriculum Study, an enterprise run by professors from the most prestigious establishment universities and funded by the National Science Foundation. Suddenly the study of evolution was in all the schools. The culture of the dominant class had triumphed, and traditional religious values, the only vestige of control that rural people had over their own lives and the lives of their families, had been taken from them. Not only in Oklahoma and Arkansas, but in California and Texas among the descendants of the Okies and Arkies of the 1930s, the new emphasis on evolution has been met by a renewed defense of the old tradition. Some of the tactics are new, "scientific creationism" for example, but the struggle is the old one. It is the struggle between the culture of a dominant class and the traditional ideology of those who feel themselves dominated. Although the declared enemy of religion, Karl Marx showed far more sympathetic understanding of the forces that generate creationism than do many modern scientists when he wrote, "Religion is the sigh of the oppressed creature, the heart of a heartless world, as it is the spirit of spiritless conditions. It is the opium of the people."

Yet, whatever our understanding of the social struggle that gives rise to creationism, whatever the desire to reconcile science and religion may be, there is no escape from the fundamental contradiction between evolution and creationism. They are irreconcilable world views. Either the world of phenomena is a consequence of the regular operation of repeatable causes and their repeatable effects, operating roughly along the lines of known physical law, or else at every instant all physical regularities may be ruptured and a totally unforeseeable set of events may occur. One must take sides on the issue of whether the sun is sure to rise tomorrow. We cannot live simultaneously in a world of natural causation and of miracles, for if one miracle can occur, there is no limit. It is for that reason that creationism cannot survive, for fundamentalists, like the rest of us, live in a world dominated by regularity. Duane Gish, no less than I, crosses seas not on foot but in machines, finds the pitcher empty when he has poured out its contents and the cupboard bare when he has eaten the last of the loaf. Creationism, in the end, is defeated by human experience.

Richard C. Lewontin
The Museum of Comparative Zoology
Harvard University
March 1982

SCIENTISTS CONFRONT CREATIONISM

1

The Word of God

BY ALICE B. KEHOE

Introduction

"It is precisely because Biblical revelation is absolutely authoritative and perspicuous that the scientific facts, rightly interpreted, will give the same testimony as that of Scripture. There is not the slightest possibility that the *facts* of science can contradict the Bible" (Morris 1974, p. 15, italics in original).

The core of the theory of scientific creationism, baldly stated, is that the Word of God has got to be true. Everything in human experience must be explained in such a way that it does not contradict the Word of God, given to us in the Judaeo-Christian Bible and translated for modern Protestant Christians under divine inspiration. Scientific observations are not more privileged than other human experiences: they, too, must conform to what we can read in the Bible. If they do not seem to conform, either they are illusions or they are erroneously interpreted. And if they are erroneously interpreted, it may be because the Devil—Satan—has caused the mistake as part of his strategy to damn mankind forever. The struggle for correct interpretations of human observations, according to many scientific creationists, is no less than the struggle between God and the Devil for the souls of humanity.

Scientific creationism is a modern version of the ageless myth of the battle between Good and Evil. It rests upon the concept that conflict is basic to the world and that everything in the world must take sides in conflicts (the principle of oppositional dualism). Either you are for God, or you are on the side of the Devil; agnosticism will not do, for any denial of knowledge of God automatically puts you in the camp of the Other. People who accept the premise that this is a world of oppositional dualism must see the Bible as either true or not true. If the Bible is true, then the message of the Gospels, that human souls can be saved through faith in Jesus Christ, is true. If the Bible is not true, then how will our souls be saved? We will be damned. This is the reasoning that makes scientific creationists committed evangelists. In their black and white world, armies are in inevitable battle. To take the Word of God as true without question can lead to glorious salvation. To question the Bible leads to eternal damnation. Love for fellow humans, Christian love, calls these Christian fundamentalists to try to persuade all men and women of this simple Truth, to win these souls in peril from the machinations of the Devil. Thus, scientific creationism can beam as a lighthouse promising a secure harbor of faith, sheltered from the horrors and complexities of today's world.

The Theory of Scientific Creationism

According to a leader of the scientific creationists, the four events that dominate the origin of life and early human history are:

> 1. Special creation of all things in six days. . . .
> 2. The curse upon all things, by which the entire cosmos was brought into a state of gradual deterioration ['entropy']. . . .
> 3. The universal Flood, which drastically changed the rates of most earth processes. . . .
> 4. The dispersion at Babel, which resulted from the sudden proliferation of languages and other cultural distinctives.
>
> [Morris 1975, p. 109]

The last event was the source of modern troubles, for it was then that "the entire monstrous complex . . . pantheism, polytheism, astrology, idolatry, mysteries, spiritism, materialism . . . was revealed to Nimrod . . . by demonic influences, perhaps by Satan himself" (Morris 1975, strata; the less intelligent and slower-moving animals (reptiles) became trapped in lower strata; and, of course, fish and other marine creatures were trapped in the lowest strata, the ancient sea. Since the receding of the deluge and the return of organisms to earth, there have been some variations within the kinds of life as creatures spread over generations throughout the earth; these adaptations might be termed microevolutionary, but they cannot be considered macroevolutionary because they represent only minor modifications; each organism remained locked within the God-given boundaries of its "kind" and no lineage of organisms changed from one into another "kind"; God's creation of the forms of life is a miracle that negates the otherwise implacable second law of thermodynamics that energy tends to become unavailable for conversion into mechanical work. So go the beliefs of the creationists.

The Background of Scientific Creationism

Scientific creationism developed from the late eighteenth-century philosophy known as Scottish Realism. This was formulated by Thomas Reid and further expounded by his disciple Dugald Stewart, both teachers in Scotland. Reid had been disturbed by conflicts within philosophy. Hume had attacked the reasoning of his predecessor, Locke. Locke, in turn, had attacked Hobbes and other seventeenth-century writers. If no eminent thinker could propose a method of reasoning that withstood criticism, then, Reid suggested, we should seek grounds for knowledge outside of pure reasoning. Reid believed he had found a method for discovering such grounds in the works of another philosopher of the seventeenth century, Francis Bacon. Bacon argued that conclusions drawn from careful observations should be the basis of scientific theories. This method of "induction" from observations seemed to Reid to promise certainties rather than the empty notions bandied about by metaphysicians. Reid, a close friend to several respected scientists in Britain, found support among them for his enthusiasm for the "scientific method." By using the scientific method of close observations, Reid hoped to discover the "first principles" upon which our world and human society are constructed. This seemed to rub out metaphysics, although in fact it only substituted one metaphysical assumption for another—that knowledge can better be gained by observing an orderly world external to the observer than by intuition or divine inspiration.

Reid and Stewart's "realist" method and philosophical theory were popular in Scotland, and among Scottish Presbyterian emigrants to America. Scottish Realism promised that hard work, following the simple formula Reid had found in Bacon, would reward the worker with a life plan for effectively dealing with the world and his fellow citizens. In their businesses and workshops, the American Presbyterians did indeed work hard, closely observing the activities of their enterprises in order to improve their design and productivity. And they prospered. An "American system" grew during the early nineteenth century, stripping products of unnecessary, expensive ornamentation so that more people could afford to buy them. This contrasted with the European practice of adding fine finishes and elaborate detail that raised the cost of products beyond the means of common people. The "American system" thus was in line with the democratization of society that characterized the United States. A philosophy that made science and the search for knowledge not much different from engineering and that put it within the reach of any ordinarily intelligent hardworking man was well suited to Americans. On it they built an economy that would in a few decades perfect mass production and give fortunes to captains of industry.

Except in New England, Presbyterians dominated the influential class of nineteenth-century America. Princeton University was a Presbyterian school. Beginning in 1837, conservatives controlled the Presbyterian General Assembly. Through Princeton and through lectures and articles in leading magazines, these conservative Presbyterians promoted a realist doctrine of knowledge alongside a firm Christian faith stressing the inerrancy of the Bible. God, they said, made the world and then created man with the mind and senses necessary to discover God's world. God in his goodness boosted man in his search for knowledge by revealing in the Bible the early history of the earth and humans and then describing how eternal salvation could be gained through faith in Jesus Christ. We can accept God's miracles and promise of redemption as true because we can confirm the truth of other parts of the Bible through observation of the natural world. Difficulty arises only when we see the other side of this coin—that if observation of the natural world does *not* confirm the Bible, then logically we have no reason to accept God's promise of salvation as true. Thus the emphasis on realism and simple unadorned logic that helped revolutionize American industry and markets posed a pitfall for the emotional assurance of many Americans. The Bible *had* to be true or all people are damned.

Creationism in the Twentieth Century

Realist ideas were not without competition. The nineteenth century saw the development of a number of concepts and scientific theories that strongly threatened them. Non-Euclidean geometries, discoveries in electricity, the wave theory of light, X-rays, and, of course, Darwin's theory of evolution through natural selection inundated educated people with facts that confounded traditional notions and the words of the Bible. Along with the scientific theories, theologians in Germany and then other countries published critiques of the books of the Bible that claimed that the Bible is not a single dictation from God but a set of compositions from different periods in Judaic history and that it contains influences from pagan thought, particularly in the Greek New Testament. This "higher criticism" seemed to many Christians a frontal attack upon their faith. The orthodox saw their hope of salvation threatened by theologians and scientists, by anarchists and socialists and nationalists who sought to deprive clergy of secular power, by freed slaves and illiterate immigrants and suffragette women who challenged the old order's leaders.

In response to the broad threats to orthodoxy, in 1910 the Presbyterian General Assembly drew up a list of "Five Fundamentals" they believed to be the rock bottom of Protestant Christianity: the miracles of Christ; his virgin birth; his bodily resurrection; his sacrifice upon the cross constituting atonement for mankind's sins; and the Bible as the directly inspired Word of God. These unique events, so contrary to everyday experience, must be accepted on faith. The last Fundamental, the inerrancy of Scripture, has implications that go far beyond the question of the salvation of one's soul. If the entire Bible is divine inspiration, then not only the core tenets of Christian salvation, but the hundreds of pages of description of historical personages, governments, battles, love affairs, and other nonmiraculous events must be precisely true, and so must the poetic myths of Genesis. Thus, while the fifth Fundamental gave certainty of the promise of redemption through Christ, it opened a pandora's box of problems because no distinction was made between spiritual messages and historical records.

Most Christians today do distinguish between the spiritual truth of the biblical passages concerning God's message of salvation and the more secular or general mythical passages. The contributions of various

Judaic and Hellenic historians over several centuries and copyists' errors are identified and sorted out from the central gospel. Allegories, parables, and poetic freedom of language are contrasted with sections that can be read as direct discourse. Scholarship is seen as an exercise of our God-given intelligence that strengthens faith by removing many of the contradictions between different books of the Bible and between common-sense experience and happenings that seem fantastic.

Among Christians who do not see "higher criticism" as destructive of faith are several respected theologians who not only accept the theory of evolution as valid but build theological insights from it. Most famous of these scholars is the late Jesuit Pierre Teilhard de Chardin, who argued that evolution is the process through which God's purpose for mankind is fulfilled. The contemporary German Jesuit Karl Rahner insists that evolution is the incarnation of God's Word, the Word made flesh, with Jesus the essence of this incarnation. Catholic thinkers who accept evolution cite two papal encyclicals, *Providentissimus Deus* and *Divino Afflante Spiritu,* supporting their view. Many Protestant theologians see evolution as a manifestation of God's plan or presence, and their ideas are echoed by several respected scientists who ponder the spiritual meaning of their work. Several hundred American scientists who are professed Christians have joined the American Scientific Affiliation, subscribing to the belief that the "Holy Scriptures are the inspired Word of God" but nevertheless able to accept evolutionary biology as a factually detailed scientific version of the poetry of Genesis. Most Christians could be labeled "theistic evolutionists": they believe God is the ultimate Creator and evolution the process through which God's creation is manifest.

Against this majority of Christians stand the scientific creationists and other "strict" creationists. Early in this century, the creationists attempted to stem the tide of ungodliness by, among other measures, forbidding the teaching of evolution in state-supported schools. This prohibition was challenged in 1925 by a young teacher in Dayton, Tennessee, John Scopes, who was egged on by local boosters who hoped to make their town famous through a trial of the validity of evolution. The county judge avoided the issue, ruling that the question at law was not the theory of evolution but whether or not Mr. Scopes had violated a state law forbidding its teaching. Scopes's guilt was not contested. He was fined $100, and the trial was over. The battery of noted scientists and liberal theologians brought to Dayton by Scopes's lawyer, Clarence Darrow, were prevented from testifying. The Tennessee law remained in effect until 1967, when it was again challenged by a high school teacher, Gary Scott. Tennessee's repeal of its requirement that creationism rather than evolution be taught in its schools was followed in 1968 by an Arkansas case over a similar law. That case reached the United States Supreme Court and there the law was found unconstitutional.

After their Supreme Court defeat in 1968, creationists changed their strategy. Led by the Institute for Creation Research in San Diego, California, many creationists campaigned to persuade school boards to mandate the teaching of both evolution and creationism. These creationists shrewdly pruned the public editions of their textbooks and pamphlets of any references to God, Jesus, or Satan. They presented "scientific creationism" as a theoretical model of the origin of life forms on a par with evolutionary biology, the alternate model. Students should be exposed to both models, they argued, and be allowed to make their own choice.

When badgered by opponents who refused to agree that "scientific creationism" was not a religious doctrine, the creationists countered that evolution was also, and even more, a religious doctrine, a property of a religion they labeled "secular Humanism." Some creationists continue to press for state (rather than local school board) requirements for teaching "scientific creationism" in public schools, and these succeeded in 1981 in convincing the Arkansas and Louisiana legislatures to pass two-model bills of a type distributed by Paul Ellwanger of Anderson, South Carolina, who insists his draft bill is compatible with federal and all state constitutions.

It is indeed ironic that in the past half century, highly conservative fundamentalists have evolved strategies better adapted to contemporary United States environments than those they employed in previous decades. By accenting the accord of the public version of their doctrine with alleged scientific facts, the creationists appeal to American respect for science and for "facts" rather than for "theory." Most Americans are realists, philosophically, with little tolerance for what appears to be wishy-washy indecision. They see little virtue in modest scientists who will talk only of validity and not of truth, who stress the difficulties of obtaining unbiased knowledge, and who speak of Heisenberg's Indeterminacy Principle as a breakthrough in modern thinking (Heisenberg said that we cannot determine with complete accuracy all the measurements we ideally need for scientific data but must work with a degree of indeterminacy). For a pragmatic American, scientific creationists' assertion that they know truth and can prove it with facts is seductive.

Some adherents of scientific creationism are not even particularly fervent Christians, but they have found the factual material provided by the creationists quite convincing. Both the Christian fundamentalist scientific creationists and their nonfundamentalist partners ignore the core principle of science, which is the observer's independence from any commitment to a preconceived idea. The modern scientist does not work blindly, but he or she must be always prepared to modify or even abandon a hypothesis that doesn't jibe with experimental or natural observations. Science uses multiple working hypotheses, choosing the one that best explains the greatest number of observations. It is directly contrary to science to declare in advance of observations that they cannot possibly contradict a preferred hypothesis such as the Genesis account of the origin of life. A fact is a generally accepted observation, not an eternally unalterable truth. (The observation that the sun goes around the earth was a fact for thousands of years, until astronomers made additional observations that would not fit this hypothesis.) The scientific creationist refuses to admit the nature of facts or the critical importance of unbiased—uncommitted—mulling over accumulated observations. Instead, he claims to know the truth by divine revelation and he arranges observations to conform to what is already written in the Bible. This prior commitment to an ancient explanation makes scientific creationism a travesty of science.

Publicly, the scientific creationists are friendly, moderate gentlemen who wax intense only when inflamed by the earnestness of their desire to enlighten those misled by the demonic theory of evolution. When they feel they can speak freely, scientific creationists may reveal a fanatic militancy. Dr. Vernon Grose, an engineer whose testimony before the California State Board of Education persuaded it in 1969 to recommend the teaching of creationism as well as evolution in science classes, explained in an interview that he could "see the forces of evil being met in these last days [before Christ's Second Coming] with an aggressive, explosive reaction of men who are led and filled by the Spirit of God . . . not simply to withstand their attack, but to attack them. . . . As C. S. Lewis has put it, we who believe in Jesus Christ are little Christs. Just as Jesus 'directly interfered in the affairs of men' by His coming to earth, so we believers today must continue to interfere with Satan's well-laid plans" (quoted by Bredesen 1971, p. 148–49).

A Protestant theologian, Robert Jewett, has criticized the strong tendency among Americans to see America as, in the phrase of the historian E. L. Tuveson (1968), the "Redeemer Nation," founded upon a vision of ideal society. American history is commonly taught as if the Jeffersonian ideals have ruled the conduct of our affairs. Textbooks play down the actual conditions of American life. One thoughtful historian identifies as characteristic of many Americans the belief that our "traditional body of ideas" cannot be at fault, and that our problems must instead flow from laxity in applying those ideals. "An extra act of allegiance and a renewed application of the original ideas" are therefore the proper recourse to rid us of "error and confusion. No philosophy could be more successfully designed to assure its own survival or to perpetuate the original tenets of eighteenth century orthodoxy," he concludes (Van Zandt 1959, p. 242). Jewett sees this belief in the absolute rightness of traditional orthodoxy as correlated with a predilection to see evil as the result of a conspiracy of outsiders, of Them against Us. This, Jewett maintains, is a mistaken view of Jesus's message. He cites chapter and verse of the Gospels to show how Jesus attempted to teach his followers to ignore this popular scapegoating and replace it with an emphasis on the power of God's goodness (1973, p. 138–39). But the notion of a demonic conspiracy dies hard. We see it played out again and again in American culture—in Westerns, in comics, in campaigns against the dirty Commies or the Wobblies or immigrants. Popeye symbolizes the American hero, taking courage "before the final confrontation with the Bad Guy," faithful to his ideals, but "curiously immune to criticism if he happens to break one of the ideals or laws in the battle. . . . His initial desire is to be passive, but when he receives the clear call to battle he must faithfully but regretfully obey." Thus, the American patriot becomes "a perfectly clean and basically passive hero, committed to lawful obedience, carrying out his highest form of faithfulness by violating cleanliness, law, and passivity" (for example, the marshall gunning down the Bad Guy on Main Street at high noon) (Jewett 1973, p. 153). Jewett's ironic insight well describes the scientific creationist.

Every human being lives in a world a little different from that of anyone else. Each of us gains from our senses slightly different impressions; each of us has had instruction and models that vary from those of our neighbors. So pluralistic a society as the United States magnifies the differences between individuals, mirrors the plurality of peoples in the world. For most of us, America's diversity and her tolerance of difference have been her greatest glory. For others, our plurality is a weakness exploited by Satan. These evangelicals believe that they know the single Truth and are called to battle Satan so that Truth may prevail.

Scientific creationists are making a mighty effort to impose their narrow doctrine upon all of us. They are well financed, well organized, and politically shrewd. They appeal to our desire for security, our wish for a simpler world. Their barrage of "facts" fitting a common-sense realist attitude can overwhelm listeners, especially when forcefully presented by men with doctoral degrees. Every citizen who values the principles upon which our republic was founded should resist the claims and simplistic appeal of the scientific creationists. The United States of America is not a Christian nation but a secular nation that offers each of us the freedom to live according to his or her own spiritual beliefs, as long as they do not involve harm to others. Any doctrinal group that wants to legislate its beliefs, whether by federal, state, or local school board authority, is un-American. Scientific creationists mean well. They wish to save our souls by promulgating Truth, but their distorted views pose a real threat to the preservation of the ideals of American society.

REFERENCES CITED

Bredesen, Harold. 1971. Anatomy of a confrontation. *Journal of the American Scientific Affiliation* 23:146–49.

Jewett, Robert. 1973. *The Captain America complex.* Philadelphia: Westminster Press.

LaHaye, Tim. 1975. Introduction. In *The troubled waters of evolution,* Henry M. Morris. San Diego: Creation-Life Pubs.

Morris, Henry M., ed. 1974. *Scientific creationism* (public school edition). San Diego: Creation-Life Pubs.

—. 1975. *The troubled waters of evolution.* San Diego: Creation-Life Pubs.

Schaeffer, Francis A. 1968. *The God who is there.* Downers Grove, Ill.: Inter-Varsity Press.

—. 1972. *He is there and he is not silent.* Wheaton, Ill.: Tyndale House.

—. 1976. *How shall we then live?* Old Tappan, N. J.: Fleming H. Revell Co.

Tuveson, Ernest Lee. 1968. *Redeemer nation.* Chicago: Univ. of Chicago Press.

Van Zandt, Roland. 1959. *The metaphysical foundations of American history.* 's-Gravenhage, The Netherlands: Mouton.

2

Scopes and Beyond: Antievolutionism and American Culture

BY JOHN R. COLE

Introduction

Organisms either evolved or they did not, whatever peoples' opinions or wishes, and the best evidence tells us that evolution is unquestionably real. Yet resistance to the idea of evolution is widespread. In 1931 biologist Julian Huxley, grandson of Darwin's great popularizer, Thomas Huxley, declared antievolutionism not just "dull, but dead" (1931, p. 185), and in 1963 Richard Hofstadter wrote, "Today the evolution controversy seems as remote as the Homeric era [to intellectuals]" (1963, p. 71). Even so, in the 1970s and 1980s legal challenges to the teaching of evolution became systematic in states as diverse as New York, California, Iowa, and Georgia, and laws were passed in Arkansas and Louisiana requiring that "creation science" be taught alongside evolution. Ronald Reagan was elected president in 1980 with the considerable help of antievolution activist groups (see Conway and Siegelman 1982). "I have a great many questions about [evolution]," he said during his campaign. "I think that recent discoveries down through the years have pointed up great flaws in it" (*Science* 1980). Antievolutionism has deep roots in American society.

The Scopes Trial

The first large confrontation between evolutionists and antievolutionists came at the Scopes trial in 1925 in Dayton, Tennessee, a trial that William Jennings Bryan said would determine whether evolution or Christianity survived. Bryan, as special prosecutor, won the case. Scopes was convicted and duly fined $100 for violating a new law forbidding the teaching of evolution. (The sentence was later overturned on a technicality by the state appellate court because the judge had set the fine, rather than the jury as the law required.

Scopes had not disputed the facts. While the school principal was ill, Scopes had filled in for him, using George William Hunter's *Civic Biology* for review purposes—a book adopted for all schools by the State Textbook Commission from 1919 to 1924; no successor had been chosen, so it was still the standard text. Scopes later wrote that he didn't remember if evolution had been discussed, but since biology was inseparable from evolution, he agreed he must have taught evolution (Scopes and Presley 1967). At the urging of local free thinkers and promoters, he reluctantly agreed to let his name be used to generate a court test of the constitutionality of the new state law banning evolution. He expected to ignite a controversy.

The trial was a bigger circus than planned. The intervention of Bryan, three-time Democratic presidential nominee, former secretary of state, and renowned orator in the cause of Christian fundamentalism, transformed a civil liberties test case into an explosive forensic contest and revival meeting. Bryan set the scene for the trial in a speech before Seventh-Day Adventists by proclaiming, "All the ills from which America suffers can be traced back to the teachings of evolution. It would be better to destroy every book ever written, and save just the first three verses of Genesis. . . ." During the trial he amplified:

> Why, my friend, if they believe [in evolution] they go back to scoff at the religion of their parents. And the parents have a right to say that no teacher paid by their money shall rob their children of faith in God and send them back to their homes skeptical, infidels, or agnostics, or atheists. [Hofstadter 1963, p. 127]

The defense, in a controversial publicity-seeking move to counter Bryan, enlisted lawyers Clarence Darrow and William Dudley Malone. Darrow, a fabled criminal lawyer, was also something of a professional agnostic and, like Malone, a skilled orator. Malone was, ironically, Bryan's former assistant in the State Department.

In fact, the issue was never joined legally. The court refused to allow a constellation of defense witnesses to testify—theologians, biologists, anthropologists, and paleontologists who came to Dayton to defend Darwinism. The issue, the judge insisted, was simply whether Scopes had taught evolution. Scopes conceded that much, although under cross-examination, the testimony of a student suggested he may have been innocent in effect, if not intent:

> DARROW: Tell me exactly what you were taught.
> WITNESS: (Long pause) That all life comes from an egg.
> DARROW: Was that all you were taught?
> WITNESS: Yes, sir.
> DARROW: Well, son, I suppose that when you heard that . . . you stopped going to church, didn't you?
> WITNESS: No, sir. [Krutch 1969, pp. 363–64]

As the trial progressed, Bryan wanted to make it a trial of Darwin versus the Bible, despite the judge's reluctance, and Darrow gleefully agreed. He led Bryan into illogical, untenable corners time and time again when Bryan insisted on testifying himself as an "expert" in biblical science. Bryan tried to prove that anyone could interpret Scripture, but he was no match for Darrow, who was coached by illustrious theologians and scientists; and although Bryan won the case, he was humiliated and mocked in the press around the world. He desisted from answering questions about the age of the earth, the antiquity of well-known archaeological sites, and so on. "I do not think about things I do not think about," said Bryan. "Do you think about things you do think about?" retorted Darrow.

Evolution emerged victorious if the debate were judged forensically rather than legally, and Bryan emerged a rather tarnished defender of the faith. He died five days later, an old statesman reduced to a laughingstock in the press, accused of leading his followers to disaster. Evolution won, so Christianity—or fundamentalism—lost.

Or so evolutionists thought. Most Americans, however, did not read the *Baltimore Sun* or *New York Times* or any other big city newspaper, or if they did it was only to confirm their worst suspicions and fears about the course and leadership of the country. Rather than ludicrous, Bryan's arguments were widely perceived as obviously true. The people of Dayton founded a small college in his name, which survives today as a bastion of antievolutionism. (For accounts of the Scopes trial, see deCamp 1968; Tompkins 1965; Ginger 1958.)

The Larger Setting of the Scopes Trial

Bryan's followers, in fact, had bigger concerns than evolutionism. The northeastern states had won the Civil War. America was changing from a nation of farmers into a nation of city dwellers and wage earners. The watershed of World War I had decimated the old empires, opening up an avenue of money and power into which much of America stepped enthusiastically. As the only industrial nation not devastated by the war, America was suddenly *the* world power—a role she had rehearsed in the Spanish-American War and in a series of Caribbean and Central American interventions. By the 1920s, rural America was outvoted, outshouted, and outfinanced, and the new city of man was a far cry from the City of God. Opposition to the coming new order had been strong among farmers at the turn of the century. They saw themselves becoming pawns in battles between railroad barons, bankers, and industrialists. William Jennings Bryan was their hero.

Nor were the Populists alone in their concerns about the future. Several factions advocating major changes aligned with them against the essentially conservative and probusiness establishment: the Progressives, who believed the emerging industrial world was rife with injustices, but that its abuses could be minimized and benefits maximized by regulating industry; the socialist and industrial labor movements, which advocated fundamental changes in the structure of economic and social power, not simply ameliorative reforms; the anarchists and nihilists; and an emerging Social Gospel movement, which opposed selfishness, competition, and inequality and advocated Christian cooperation in the name of either socialism or progressivism.

Most of these movements were primarily urban and industrial; Populists tended to be rural and agrarian. In the early 1900s, they were the most powerful of the dissenters from the status quo; but they lacked any coherent theoretical analysis of social ills and in the long run could not sustain themselves as a political force.

On the other hand, from its beginning, the urban Left raised many of the same criticisms of society as the Populists, but did not argue as the Populists had that technology could be swept away. To the contrary, orthodox Marxists pinned their hopes for revolution on an industrial proletariat. The Left survived when populism faded, in large part because leftism was "modernist," and it often cultivated rather than ridiculed intellectuals who could provide social analyses instead of simply making emotional appeals for "fairness." The political wellspring that nurtured Populists, Progressives and Leftists could sustain only the latter two groups when the rural population and economy continued to erode. Disavowing cities and factories, the Populists lost their reasons to exist as a political force when industrialization triumphed. As a cultural force, however, *alienation,* the basic reason for the existence of populism, continued: alienation from politics was added to alienation from intellectual and economic currents. Populist movements continued to spring up with fairly little consistency except in their reliance upon emotional rather than intellectual appeal, and over the years ceased to be dominated by farmers advocating wealth redistribution and opposing big business. In the process, much of the *old* populism that had been unable to cope with changing times disappeared, allowing the absorption of new constituents more comfortable with the emerging technological world.

"Progressives" believed in the inevitability and desirability of progress and established themselves in both the Republican and Democratic parties, as well as in their own, more radical parties. Neither mainstream party was revolutionary, and both sought to channel potentially explosive dissent into ameliorative reform. Unsuccessful presidential candidate Al Smith and then, spectacularly, Franklin Roosevelt, brought many progressive and populist goals into the fabric of Democratic philosophy without challenging the basic social order. The progressive wing of the Republican Party faded in importance after the First World War, as did the separate progressive parties.

But populism was not as interested in expanding its base as was the Democratic party. Populists might sympathize with the plight of industrial workers; but rather than make common cause with Progressives and the struggling union movement, they tended to deplore the industrial world and to preach a return to the simple days of the yeoman farmer and the craftsman. The Populists were thus a revitalization movement harking back to a golden past (cf., Wallace 1968; Glad 1960). Populists were "anti-intellectual" in effect, intentionally or not, because they opposed "progress" that was rending their previously comfortable (or at least familiar) social and economic fabric; industrial technology was the demon, educated people invented and ran it, and evolutionists, they thought, defended it. "Too much" education was often seen as wasteful and morally questionable, except in practical fields like medicine, agriculture, or law. Thus, by the time of the Scopes trial, populism had lost its national vitality. Its activism was restricted to religion where God gave comfort as the political and economic system had not. From political revival with religious overtones had sprung religious revival with political overtones, its secular power dormant.

Darwinism, Spencerism, Socialism, and Antievolutionism

Darwin's theory had been in partial eclipse at the turn of the century because of supposed conflicts with newly rediscovered Mendelian genetics, but evolution nevertheless was embraced by a wide range of scientists, scholars, reformers, theologians, and tycoons—including people who came to be known as "Social Darwinists," followers of Darwin's contemporary, Herbert Spencer. "Social Darwinism" was Spencer's attempt to synthesize biology, physics, sociology, and philosophy. It explained that whatever existed was "natural"—the rich were rich and the poor were poor because of "natural law." Spencer's explanation of society's class structure as "natural" pleased the robber barons; many conservatives embraced evolution because they thought it showed natural change to be glacially slow and nonrevolutionary. Ironically, Spencerism fit indirectly into Marx's view, as well, although Marx believed the proletariat would "naturally" come to deserved power some day.

To Populists opposing change and longing for a lost past, however, the naturalness or inevitability of change was an abhorent doctrine. Social Darwinism rationalized a system they hated, and Darwinism itself was threatening because it claimed to prove that all in nature was flux, outside human control. The Bible's assurance that humans had dominion over nature was challenged. Supposed eternal verities such as the plants and the animals were not stable through time. Most of all, Darwinism matter of factly showed humans to be animals and not the centerpiece or epitome of Creation. If Man was made in God's image, was God an animal? Populism suggested that farmers had been driven out of their idyllic fields like Adam and Eve from the Garden of Eden; what would it mean if there had been no Adam or Eve or if they had been lower animals rather than near-angels? Was there no superior past to which humanity could return? Was the Garden populated only by microorganisms? Was the personal God to be replaced forever by a religion totally free of supernaturalism and based upon human potential as evolutionist Julian Huxley predicted in "Science and the Future of Religion"?

> Gone is the bearded Jehovah, gone is Milton's conversational God the Father, and in their place are creative first principles, emmanent spirit, divine purposes informing the slow movement of evolutionary progress, and so forth. [1931, pp. 235–36]

A blunt worry was expressed by the famous Scopes era antievolutionist George McCready Price: "No Adam, no fall; no fall, no atonement; no atonement, no Savior" (Furniss 1954, p. 16). At the same time, Presbyterian Albert Johnson claimed that evolution leads "to sensuality, carnality, Bolshevism and the Red Flag" (Gatewood 1969, p. 24). A Louisiana clergyman wrote,

> A modernist in government is an anarchist and Bolshevik; in science he is an evolutionist; in business he is a Communist; in art a futurist; in music his name is jazz and in religion an atheist and infidel. [Gatewood 1969, p. 6]

Despite such views, many religious leaders in the late nineteenth and early twentieth centuries were friendly to evolution. When *The Origin of Species* was published, most religious scholars in America and Europe felt it was a threat to their beliefs, but soon leading New England and northeastern clerics were praising evolution as a remarkable proof of God's wisdom and hailing Darwin as a theist and virtual theologian.

Henry Ward Beecher and other leaders of the nineteenth-century American religious establishment especially embraced Herbert Spencer's evolutionism. The natural order of things was God's will, they argued.

Yale's William Graham Sumner summarized the "proper" intellectual position of the time: "Let it be understood that we cannot go outside of this alternative: liberty, *inequality,* survival of the fittest; not liberty, *equality,* survival of the unfittest" (Hofstadter 1955, p. 51, italics added). In practical terms, he said this meant: "The millionaires are a product of natural selection, acting on the whole body of men to pick out those who can meet the requirements of certain work to be done. . . . It is because they are thus selected that wealth—both their own and that entrusted to them—aggregates under their hands" (Hofstadter 1955, p. 58). This was obviously welcome news to the rich but not to the poor, who were disinherited now by God as well as by an often rapacious economic system. It generated the colorful phenomenon of industrial barons claiming to be philosopher kings, writing books and articles to celebrate the congruence of God's will, natural law, and their own prosperity. In 1900 John D. Rockefeller epitomized their pious claims to virtue: "The growth of a large business is merely survival of the fittest. . . . [forcing small companies out of business] is not an evil tendency in business. It is merely the working-out of a law of nature and a law of God" (Hofstadter 1955, p. 45). Fifteen years later he wrote, "God gave me my money" (Appleman 1979, p. 387).

Darwin, or really Spencer's Social Darwinism, accelerated a sort of rebirth of the old idea of the divine right of kings, proving it scientifically for the age of technology. Of course, people who oppose the scientifically intellectualized social order sometimes become explicitly antiscientific. Told that both God and Darwin decreed popular misery and that evolutionary law decreed the direction in which they were "naturally" headed, many people resented and opposed the new doctrines of inequality. They thus rejected Darwin and turned to churches other than those swept up in Social Darwinism. Darwin's biology was not necessarily the issue to many critics except to the extent that Social Darwinists (certainly including many biologists) misused it to rationalize as "natural progress" the headlong rush into brutal industrial society. Antievolutionist analyses of social and political ills were naïve in their understanding of the processes of history when they believed that the solution was to return to a mythical golden age. But they accurately perceived that their troubles were caused by other people rather than by the immutable will of God or nature.

To the populist anti-Spencerians, God was on their side, and evolutionism was a sin. Times and conditions might change, but the sense of security afforded by deeply held beliefs would not lose its appeal. The populist tradition, which once embraced rather radical political positions, came gradually to disavow much of its secular program. After World War I the ethos of "progress" and business prosperity dominated American politics. While the Great Depression severely taxed the myth of inevitable material and social progress, and the political reaction to it included adoption of ameliorative policies echoing more radical political platforms, the basic ordering of society survived. Wealth and power accrued to a smaller and smaller percentage of the population, continuing the trend that had begun at the turn of the century during the early populist days. World War II and the postwar boom furthered the trend toward centralization of economic and political power and social integration.

But as America industrialized, urbanized, and culturally homogenized, the old populist idea that everyone (or white Protestants, at least) should totally control his own destiny—that everyone's opinion is equal—did not slacken. To the contrary, American ideology reinforced the ideal of individualistic independence from authorities and material constraints: The Horatio Alger stories of the nineteenth century reflected populist beliefs that everyone who wants to *can* succeed—that average people have the skills to cope with problems without elite education. However, an increasingly technical world emerged in the twentieth century, and it was less and less possible to be a yeoman farmer. The South and Midwest today are as dependent upon high technology as any other part of the country, and even farming has become an industrial operation. Many people still share Populists' views and sense of alienation, but they are more likely to be wage earners in electronics companies than independent farmers or craftsmen.

It may seem paradoxical that archconservative politics is today often associated with antievolutionism and that inheritors of the populist tradition now often endorse the New Right political cause, given the earlier conservative establishment's fondness for Spencerism. The old Populists who opposed World War I because they saw in it the results of belief in progress and technology would be amazed to see their antievolutionist conclusions preached from "electronic pulpits" via earth satellite television relays and accompanied by the message that MX missiles and unbridled capitalism are God's will (cf., Fitzgerald

1981). 'False consciousness', or the misperception of basic issues and forces by many people, is a prominent feature of culture. It is especially cultivated in complex, stratified society where some people profit from the divisive misunderstandings of others about how things work. Populists past and present exemplify a range of atheoretical attributions of discontent whose antiestablishmentarianism can be perceptive, irrational, and bigoted at the same time. But activist right wing religious movements such as the "Moral Majority" and the political sophistication of allegedly nonpartisan groups such as the Institute for Creation Research have proven to be potent lobbies for religious and social conservatism. Their success belies the liberals' myth that the Scopes trial settled the issues of evolution, education, and the value of intellectualism.

The Textbook Barometer

Scopes proved to be the only person ever tried for violating the "Monkey Law" in Tennessee, and laws in other states were similarly unenforced in following decades. The derisive trial publicity cast creationism in such a foolish light, it is easy to conclude that the Tennessee and other states' antievolution laws failed long before they were ruled unconstitutional in the late 1960s. Was creation legislation a harmless nod to a vocal constituency, since people violating these laws were not prosecuted? Unfortunately, we cannot measure the extent to which teachers who taught or wanted to teach evolution were fired, harrassed, or intimidated by pressure groups using the laws as formal justification, let alone how often state endorsement of creationism influenced people more subtly. We do, however, have an excellent barometer of public exposure to evolution in the classroom: textbook content.

Nineteen twenty-five was a watershed year. From that point on textbooks tended to remove or dilute their treatments of evolution; some publishers began the trend in 1924, just as antievolution laws began to proliferate in southern states. Publishers seem to have viewed the trial as a warning of the need for self-censorship to avoid loss of sales, not as a civil liberties victory for evolutionism. A number of texts that were previously outspoken about the importance of evolution as the basis of modern biology downplayed the theme in their new editions. The few new texts that openly discussed evolution, such as Alfred Kinsey's *Intro-*

duction to Biology (1926), were not adopted widely. The absence of controversial material, rather than scientific quality, determined book adoption by many school committees, and antievolutionists had learned to bring effective pressure on them.

A number of states have statewide book selection that admits or excludes publishers from a large market, sometimes for years. When the state is the size of Texas or California, a significant portion of a publisher's national sales may depend upon a very small number of adoption decisions. In the 1970s socially and religiously conservative southern states had roughly half the nation's total high school enrollment in biology classes despite having only a third or fewer of the total American high school students; only about one in six northerners studied biology while one in four did so in the South, perhaps because of the greater importance of agriculture and animal husbandry there (Grabiner and Miller 1974). Furthermore, most statewide book committees were in the South, and *none* were in the East. It made economic sense for publishers to acknowledge the power of such a large block of their markets. Thus the 1970s and early 1980s witnessed changes in secondary school biology texts, much as did the period immediately following the Scopes trial. In the late 1920s and 1930s, publishers played it safe, bowing to the demands of antievolutionists, both because there were many of them and because there was no counterbalancing lobby from evolutionists. Evolutionists were fairly quiet after their Scopes "victory," while the more astute creationists adhered to the political adage, "Don't get mad —get even."

Few precollege students learned much about evolution, before or after the trial, beyond the fact that biologists accepted and believed it. Frequently, evolution was presented to students and the public as a belief system rather than as a scientific theory that explains specific data. Even today, few nonspecialists are aware of the data being explained by evolutionary theory, nor are they aware that many alternative evolutionary explanations have been and are being proposed and tested (Godfrey 1980).

Pollsters ask if people "believe in" evolution or the Bible, suggesting that a choice between these two alternatives is necessary and that the issue *is* belief. Julian Huxley predicted that "evolutionary humanism" was to be mankind's next "religion" (1957), and countless neo-Darwinians have echoed his sentiments less dramatically. Today's "scientific creationists" sometimes sieze upon such rhetoric as proof that evolution is in fact a religion rather than a science and a belief system diametrically opposed to belief in God as well.

Until the 1950s professional evolutionists were fairly content to teach their intellectual subject to other intellectuals, offended by literal book-bannings, perhaps, but fairly unconcerned about public school biology being generally non- if not antievolutionist. In 1957 the Russian launching of Sputnik was a rude awakening to intellectuals and everyone else in America. The political, economic, and military establishment panicked at the thought that Russia seemed to be ahead of America in the "science race," and the result was a large-scale reappraisal of American education that resulted in a massive federal commitment of money and attention to science education. (Ironically, in 1957 the Soviet Union was only beginning to officially accept neo-Mendelian genetics, having until then followed the Lamarckian precepts of Trofim Lysenko.) The cold war might have thawed slightly since its peak (Stalin was dead and McCarthyite hysteria had subsided), but fear of the Communists proved more persuasive than fear of ignorance or enthusiasm for science education in the abstract. Out of the cold war grew a series of programs whose nonintellectual genesis was epitomized by the very titles of the National Defense Education Act and the National Defense Foreign Languages Act.

One of the new forces was the Biological Sciences Curriculum Study (BSCS), created in 1959 to develop a thoroughly reformed biology curriculum for the nation's schools. Its textbooks on cellular biology, ecology, and molecular analysis appeared in 1963–64 as completely fresh and totally evolutionary introductions to biology for high schools. The committee drew on the best professional science available rather than on the consensus-oriented, bland nonevolutionism of most previous textbooks. (For example, one of the most popular *non*-BSCS texts, by Moon, Mann, and Otto 1957, treated evolution only at the end of the book and used the odd term "racial development" rather than evolution.) The BSCS books quickly became popular and were adopted by nearly half of American high schools by 1970. Professional educators liked them, but some parents liked neither the books nor professional educators. After several emotional debates, two of the three BSCS titles once adopted in Texas were dropped from the "approved list" in 1969. A similar drop off in BSCS use occurred in other states when textbook vigilantes raised objections.

In 1980 creation instruction was officially available in states such as Wisconsin, Missouri, and South Dakota (Gorman 1980), and in other states a *de facto* "equal time" formula prevailed in many school districts (Weinberg 1980); in 1981 the Arkansas and Louisiana legislatures passed "equal time" laws. Texas dictated in 1980 that evolution be presented as "only one of several explanations of the origin of mankind" (Gorman 1980, p. 94), and in 1981 a California court ruled that evolution be taught as a "theory" rather than as a "fact." Lois Arnold, senior science editor at Prentice-Hall, said, "We don't advocate the idea of scientific creation, but we felt we had to represent other points of view," and another editor whose book presents creationism said, "After all we are in the business of selling textbooks in the 1980s" (Gorman 1980, p. 94). The resurgence of creationism in major textbooks in the 1980s is a reaction to political and economic pressures, not a changing scientific evaluation of evolution.

The clearest example of the political nature of contemporary antievolutionism, however, is not the BSCS project but the MACOS project. *Man: A Course of Study,* begun in 1963 by the National Science Foundation, was finally published in 1970 by the Education Development Center as an introduction to evolution and behavioral/social science for elementary school students, usually those in the fifth and sixth grades. In 1980 Ronald Reagan used MACOS in his successful presidential campaign as an example of the federal government endorsing subversive values, and he asked why NSF did not instead develop curriculums supporting Christian values (*Science* 1980).

MACOS books and filmstrips comprised a rather complex, expensive package for school use, and no doubt many schools did not adopt the program for simple budgetary reasons. No commercial publisher would touch the project because "religious groups would not endorse the teaching of this type of material," according to a spokesman for the small foundation that agreed to undertake it (Nelkin 1977, p. 34). By 1974, seventeen hundred school districts in forty-seven states had adopted MACOS, but in 1975, when organized opposition began to assert itself, this sales rate plummeted 70 percent.

MACOS asked students to study an animal (i.e., salmon) and another culture (i.e., the Netsilik Eskimos) and to compare the lives of the animal and of the people in the other culture with their own lives, focusing on questions like what is human about humans? how did we get this way? what are our options for the future? Animal research, ethnography, and self-study were all part of the course. The combination proved explosive. Parents reacted in force:

> "I will never say I came from an ape."
> "Teaching that man is an animal and nothing more is denying the existence of God and Religion."
> "I wonder how many parents would be happy to see their son identify with a baboon instead of his father?"
> "[MACOS] will break down the moral fiber of American youth."
> "The education experts are dictating our values."
> "It eliminates the beliefs, values, and allegiances of children, alienating them from their parents." [Nelkin 1977, p. 108–9]

Right-wing organizations and religious crusaders worked together in organized campaigns to reverse previous MACOS course adoption decisions and to prevent new ones. Texans Mel and Norma Gabler, long successful forces in antievolution and antipornography crusades, joined the fray. Mrs. Gabler was influential in organizing parents as far away as Queensland, Australia, to ban MACOS. One study of the anti-MACOS campaign demonstrated that its participants were "absolutist," "didactic," "totalitarian," and "reactionary," while the MACOS advocates were the opposite: "relativistic," "tolerant," "secular" (Smith and Knight 1978). But that was the very point: the two sides in the dispute did not share the same ideals for education at all. Cultural relativists who supported both MACOS and the right of anti-MACOS people to disagree were ineffective lobbyists compared with their absolutist foes.

The National Science Foundation, which had provided $4.8 million to develop MACOS, suddenly was attacked in Congress. The House of Representatives passed the Bauman Amendment in 1975, giving Congress direct supervision and veto power over every single NSF research grant, but the bill died in the Senate. Representative John Conlan of Arizona led the attack on policies favoring what he called "low priority behavioral research and curriculum projects" rather than funding to create practical jobs in private industry (Nelkin 1977, p. 119). That same year NSF split off its troublesome biological and social science work into a separate directorate, and in 1980 a proposal was discussed to divide biological from social sciences in the foundation. Nineteen eighty-one saw a drastic cut in federal support for social science research, and science *education* was virtually eliminated from the federal budget.

Cultural relativism, central to MACOS, is the idea that one culture is not superior to and should not judge others. It may be the single most influential anthropological precept. It is easily related to liberal and libertarian doctrines of personal, racial, and political equality and even anti-imperialism, and anthropologists have given it scientific as well as moral credence. The Bible as well admonishes "judge not, lest ye be judged," leaving absolutism in the hands of God. Like evolution, relativism implies that what some view as absolutes are actually parts of continuous variation. Absolutes or a priori "givens" are not assumed in modern science. To relativists, Western society, American politics and capitalism, and Judaeo-Christian ideas of morality are not absolute or perfect any more than is New Guinea tribal life. To people committed to absolute standards defined by the will of God (*or* nature), relativism is a humbling, subversive doctrine. It removes an individual's group from the pinnacle of culture, just as evolution's demonstration that people are simply one more variety of animal removes humans from the center of life, and just as the discoveries of Galileo and Copernicus earlier removed the earth from its central role in the universe.

"Humanism," another red flag term to the New Right, is closely related to relativism and evolution. Roughly defined, it is a belief that people, not gods, must be in charge of solving human problems. It relates natural law to human behavior much the way evolution does (cf., *Newsweek* 1981). Even some orthodox or main-line churches worry about Humanism detracting from supernaturalism; "salvation by works" versus divine election is an age-old Christian debate. A 1980 American Roman Catholic Bishops' pastoral letter expressed concern about the humanistic element of Marxism: "Marxist transcendence . . . remains within the scope of human attainment. Christian transcendence consists in being assumed into an order totally beyond the reach of human endeavor" (*Des Moines Register* 1980). While the Roman Catholic church does not attack evolution, arguments such as the above are crucial to antievolutionists, as well.

In its fight against MACOS the New Right charged that "secular Humanism," evolutionism, and cultural relativism were elements of a conspiracy to subvert students, substituting relativism for beliefs in nationalism, old time religion, and the natural authority of leaders, parents and traditional values. By asking students to question authorities and to discuss rather than simply memorize values, MACOS epitomized the antievolutionists' fears of the social implications of evolution. MACOS organizers were impolitic not to have foreseen these negative reactions and planned accordingly. Instead, they sometimes used arrogant language and tactics that enraged the opposition:

> It will not do to dream nostalgically of simpler times when children presumably grew up believing in the love of God, the virtue of hard work, the sanctity of the family, and the nobility of the Western historical tradition. . . . We must understand . . . what causes . . . these things. [Dow 1975, p. 81]

By objective standards, the MACOS program did not advocate atheism, socialism, communism, immorality, or family dissolution. It simply asked students to make guided judgments about behavior without being *told* that the behaviors their parents believed in were the only right ones. Such liberalism is a far cry from the leftism MACOS critics feared. But MACOS struck nerves because it was designed for young children. Parents who oppose evolution might believe their children could stand exposure to it in elective high school biology courses but not in required elementary school science classes. As in arguments over sex education, it is difficult for school administrators to argue against parental control over what children learn, whatever the experts might prove they "need" to learn.

At the Scopes trial and in the MACOS debate, the experts asserted in vain that students should learn what is necessary to be "citizen scientists" able to cope with a world filled with problems that science could solve. But this was the epitome of the humanist position—that people rather than gods or authorities were humanity's best hope. Antievolutionists and other conservatives, from the John Birch Society and the Heritage Foundation to the founders of "Christian Academies," fought this idea as state interference with parental rights. "The idea that an individual should collect evidence and decide for himself is anathema [to the fundamentalist New Right]" (Fitzgerald 1981, p. 99). Reading, writing, and arithmetic are noncontroversial, and sciences such as physics and chemistry, despite their potentially harmful applications, are also seen as value-free and therefore safe. But some parents clearly do not trust their children to make their own judgements after learning methods of inquiry; they prefer schools to give pat answers rather than reasoning skills, and they especially distrust "impractical" intellectuals who claim to know what is best for their children, especially when their expertise is often approximate and when they claim to take a neutral stance on the very topics some parents want authoritatively defined. The teaching of methods and theories, which are the heart of science, is scorned because it is potentially subversive: who knows where questioning will lead? John Dewey wrote enthusiastically:

> If we once start thinking no one can guarantee what will be the outcome, except that many objects, industries and institutions will be surely doomed. Every thinker puts some portion of an apparently stable world in peril, and no one can wholly predict what will emerge in its place. [1929, p. 1]

But what is here a virtue to Dewey is a threat to many others.

Antievolutionism and Anti-intellectualism

For about a century, America's dominant culture has prided itself on living in an age of science and technology. It has become more and more necessary for individuals to accept the virtues of modernism, progress, and change to fit into "proper" society. Science may not be worshipped overtly, but technology generally is (cf., Cole 1980, Etzioni and Nunn 1974), and many of science's assumptions and ideas are taken for granted by anyone who wants to be identified as educated or middle-class. Electricity, nuclear power, or breeding hybrid roses may not be understood, but as a political act people may choose to think they *should* understand them to avoid appearing ignorant. Conversely, to reject major elements of modernism is also a political act. Sincerely or cynically, and often ambivalently, attacking an intellectual or scientific doctrine has been popular from the evangelism of colonial times to Senator William Proxmire's Golden Fleece Awards. Attacking evolutionism may stem from a simple desire to attack the establishment and to express general discontent rather than from a straightforward disagreement with a biological theory.

Populism is often hailed as unalloyed anti-intellectualism—part of a long American tradition traceable back as far as the early eighteenth-century Great Awakening and its call to abandon rationality in favor of revelation. Richard Hofstadter's *Anti-intellectualism in American Life* (1963) chronicles this pervasive theme. The intellectual tradition he defines is based upon people living *for* ideas and analysis rather than simply *using* ideas, as he claims Edison used ideas in physics and chemistry, for example. The intellectual is one who turns answers into questions, he writes (1963, p. 25–30). But by these standards some antievolutionists and Populists would seem to qualify as intellectuals. Princeton, Brown, Rutgers, and Dartmouth, for example, were founded by evangelicals in reaction against the intellectual establishment.

Hofstadter's book inadvertently documents an intellectual tradition among people whose leaders wrote and spoke eloquently of the virtues of ignorance. But most anti-intellectuals were not eloquent. The words of a Georgia legislator fifty years ago illustrate the worst of this tradition:

> Read the Bible. It teaches you how to act. Read the hymn book. It contains the finest poetry ever written. Read the almanac. It shows you how to figure out what the weather will be. There isn't another book that it is necessary for anyone to read, and therefore I am opposed to all libraries. . . . [Hofstadter 1963, p. 125]

More recently the Reverend Jerry Falwell, leader of the Moral Majority organization, warned his followers not to read books other than the Bible (Fitzgerald 1981, p. 99).

Hofstadter's definition of intellectualism is well documented, widely accepted, and restrictive. Implying that a Thomas Edison or an Alexander Graham Bell was not really an intellectual perpetuates the false opposition between the intellectual but impractical professor and the nonintellectual but practical "regular guy." "Practical" professors understandably resent this popular caricature of the intellectual. But few would challenge Hofstadter's basic premise that anti-intellectualism exists and is a basic theme in American history. Whether or not they conform to Hofstadter's definitions specifically, significant numbers of people have actively opposed or at least resented the "intellectual class." Daniel Boone and Davy Crockett are more typical American folk heroes than André Malraux, Henry David Thoreau, or Goethe.

Anti-intellectualism is seen as a virtue among anti-intellectuals, apparently to the surprise of the intelligentsia. Some anti-intellectualism has been expressly humane and nurturing of values that aid in practical survival. Many intellectuals today would agree that Populists were correct in rejecting Social Darwinist dogma, though their reasons for doing so were unscientific. To the extent modern creationism gives comfort to people, scientists might do well not to condemn it, but when creationism interferes with the education of nonbelievers through censorship, curriculum changes, or other political acts, the situation is different. By advocating antiscientific beliefs in an age of science, or by promoting an authority-based version of science, creationists contribute to the kind of ineffective education that led to the Sputnik shock.

Antievolutionism is best understood as an aspect of the anti-intellectual tradition, but it has varied through time, as has intellectualism. In retrospect neither the antievolutionists nor the evolutionists have a monopoly on virtue. But to people who believe science can and should have positive value to society, the occasional virtues of antievolutionists must be seen as accidents in the midst of a tradition glorifying noncritical acceptance of authority. The errors of scientists have been committed within a system devoted to self-analysis, testing, and self-correction rather than to acceptance of the heavy hand of tradition. Today's antievolutionists do not oppose science formally, but in seeming to endorse science while rejecting basic aspects of it, they foster a schizophrenic approach to the empirical world. It is ironic that the early twentieth-century Populists were more intellectually consistent than this, and this irony is compounded by many intellectuals' smug, nonanalytical reaction to the Scopes trial, the event that symbolically ended the populist era and prepared the ground for a new fundamentalism of technocrats and suburbanites doing similar things more efficiently and stripped of many populist virtues.

REFERENCES CITED

Appleman, Philip. 1979. *Darwin.* 2nd ed. New York: Norton.

Cole, John R. 1980. Cult archaeology and unscientific method and theory. In *Advances in archaeological method and theory*, ed. M. Schiffer, pp. 1–33. New York: Academic Press.

Conway, Flo, and Siegelman, Jim. 1982. *Holy terror.* Garden City, N. Y.: Doubleday.

DeCamp, L. Sprague. 1968. *The great monkey trial.* Garden City, N.Y.: Doubleday.

Des Moines Register. 1980. U. S. bishops vote to drop 'sexist' words in prayers. (13 Nov.) p. 1.

Dewey, John. 1929. *Characters and events,* ed. Joseph Ratner. New York: H. Holt Co.

Dow, Peter. 1975. MACOS: the study of human behavior as one road to survival. *Phi Delta Kappan* 57: 81ff.

Etzioni, Amitai, and Nunn, Clyde. 1974. The public appreciation of science in contemporary America. *Daedalus* 103: 191–205.

Fitzgerald, Frances. 1981. A disciplined, charging army. *New Yorker* (18 May), pp. 53–141.

Gatewood, Willard B., Jr., ed. 1969. *Controversy in the twenties: fundamentalism, modernism and evolution.* Nashville: Vanderbilt Univ. Press.

Ginger, Ray. 1958. *Six days or forever? Tennessee vs. John Thomas Scopes.* Boston: Beacon Press.

Glad, Paul W. 1960. *The trumpet soundeth: William Jennings Bryan and his democracy, 1891–1912.* Lincoln: Univ. of Nebraska Press.

Godfrey, Laurie R. 1980. The misunderstanding of evolutionary biology. *The Skeptical Inquirer,* vol. 4, no. 4, pp. 69–73.

Gorman, James. 1980. Creationism on the rise. *Discover* (October), pp. 92–94.

Grabiner, Judith, and Miller, Peter. 1974. Effects of the Scopes trial. *Science* 185:832–35.

Hofstadter, Richard. 1955. *Social Darwinism in American thought.* Philadelphia: Univ. of Pennsylvania Press.

—. 1963. *Anti-intellectualism in American life.* New York: Knopf.

Huxley, Julian S. 1931. *What dare I think? The challenge of modern science to human action and belief.* New York: Harper and Bros.

—. 1957. *Religion without revelation.* New York: New American Library.

Krutch, Joseph Wood. 1969. Dayton: then and now. In *Controversy in the twenties,* ed. W. Gatewood, pp. 358–67. Nashville, Tenn.: Vanderbilt Univ. Press.

Nelkin, Dorothy. 1977. *Science textbook controversies and the politics of equal time.* Cambridge, Mass.: MIT Press.

Newsweek 1981. The Right's new bogeyman. (6 July), pp. 48–50.

Science. 1980. Republican candidate picks fight with Darwin. 209:1214.

Scopes, John T., and Presley, James. 1967. *Center of the storm: memoirs of John T. Scopes.* New York: Holt, Rinehart and Winston.

Smith, Richard, and Knight, John. 1978. The politics of educational knowledge: a case study. Paper read at meeting of Sociological Association of Australia and New Zealand, 20 May 1978.

Tompkins, Jerry R., ed. 1965. *D-days at Dayton.* Baton Rouge,: La. LSU Press.

Wallace, Anthony F. C. 1966. *Religion: an anthropological view.* New York: Random House.

Weinberg, Stanley L. 1980. Reactions to creationism in Iowa. *Creation/Evolution,* vol. 1, no. 2, pp. 1–7.

3

The Ages of the Earth and the Universe

BY GEORGE O. ABELL

> To ask or search I blame thee not, for heav'n
> Is as the Book of God before thee set,
> Wherein to read his wondrous Works, and learne
> His Seasons, Hours, or Days, or Months, or Yeares.
>
> Raphael to Adam and Eve,
> *Paradise Lost,* Book VIII.

In the years 1650 to 1654, the Irish archbishop James Ussher wrote his *Annales Veteris et Nove Testamenti,* in which he set out his famous biblical chronology and his conclusion that the universe was created in 4004 B.C. Perhaps because of Ussher's eminence, Christian theologians adopted his chronology as a timetable for (or explanation of) world history over those of hundreds of competitors. Based on a literal interpretation of the absolute correctness of the Bible (or a translation thereof), Ussher's timetable of events (with some minor modifications) is the basis of the fundamentalist creationism view that the earth is only a few thousand years old.

Science cannot prove that that view is wrong.

Science cannot *prove* anything, except on the assumption of certain postulates or axioms. For example, if we accept the correctness of gravitational theory (either the Newtonian or Einsteinian version), which has, after all, been impressively successful in predicting new planets and in guiding the Voyager spacecraft through and about the rings and satellites of Saturn, then it follows logically that the earth must revolve about the sun, and not vice versa. But perhaps some other theory can be found that will allow all known gravitational phenomena and yet have the earth at rest. It is not hard to think up such a theory: the reality of the universe might all be a dream, for example.

Similarly, one can think up an easy explanation for all of the phenomena described below that point to an old age for the earth and universe: namely, all was created in 4004 B.C. (or thereabouts) complete with the "appearance" of great age, perhaps partly decayed radioactive elements, and light well on the way to earth from remote stars. It would be as if God had perpetrated a gigantic hoax on us, perhaps to test our faith in his Word (or Ussher's interpretation of it). Science has no refutation of that theory; nor can it prove that the universe was not created at 10:00 P.M. last night with our memories intact.

But such ideas are not really theories; they are untestable assertions. Perhaps the creationists' assertions are true (and many people accept them as such), but they are *not* part of science. I suspect that they, like other untestable assertions and speculations, will lead to no new discoveries and will shed no new light on ongoing questions since by their very nature they are unverifiable.

Science, on the other hand, while not providing absolute truths, does find systematic order and develops models that allow us to understand nature and her behavior. Moreover, the method of science—observation, hypothesis, and test—leads to new knowledge and deeper understanding.

Most scientists take for granted that these models, at least the extremely well tested ones, do describe reality—that is, that the earth really is round, does turn on its axis, and does go around the sun. This acceptance is really a religious one, and, if challenged, those same scientists will usually agree that technically no truth can be proven by science to be absolute. Nevertheless, it provides ways for us to understand things in a rational manner, and exploitation of that understanding leads to a technology and to greater comforts and enjoyment in life.

Science is an active discipline. There is always a frontier beyond which we have yet no understanding and at which our understanding is incomplete. At and near the frontier our models are subject to modification, perhaps rejection, as new information becomes available. But the existence of uncertainty at the frontier does not negate our knowledge in the well-trodden foreground. Whether or not neutrinos turn out to have finite rest mass, for example, has no bearing on the motion of the earth about the sun. All this said, what have we learned from science about the age of the world?

The Oldest Trees

In temperate climates the rate of growth of new wood in the trunk of a tree varies with the seasons, so that during a given year there are periods of rapid and slow development. The annual variation in growth rate results in a series of concentric rings in certain trees that can be seen in the cross section of a trunk, one ring for each year of growth. A count of the rings, therefore, gives the age of the tree. Moreover, variations from year to year result in rings of different widths. Inspection of the pattern of wide and narrow rings makes it possible to tie in the chronology of a tree just felled with that discerned from the trunk of a much older tree that might have been lying in the vicinity since it fell hundreds of years earlier.

The oldest known trees are the bristlecone pines, small, gnarled trees that grow in the mountains of California. Some living bristlecone pines are more than forty-five hundred years old. But dead ones have been found that are much older yet, and by matching rings in old and recently felled bristlecones a continuous chronology for these trees has been traced back for approximately eight thousand years—already much longer than the age of the earth according to Bishop Ussher.

Continental Drift

In 1912 the German meteorologist Alfred Wegener (1880–1930) noted the similarity between the coast line of eastern North and South America and that of western Europe and Africa. The fit is actually even better for the continental shelves. Moreover, there is a continuity of geological features in the corresponding places in Africa and South America and in Europe and North America. This was pretty good circumstantial evidence that the Americas are gradually drifting away from Europe and Africa, but the idea was not seriously considered until the 1960s, when positive evidence was found that the mid-Atlantic sea floor is spreading, pushing the continents apart.

The drift in the continents is produced by *plate tectonics.* The entire surface of the earth, down to a depth of some thirty to seventy miles, consists of a horizontal mosaic of about ten major plates, and some minor ones, somewhat like a cracked shell on a hardboiled egg. The plates are slowly circulating over the globe because of an extremely slow convection in the solid rock of the underlying mantle of the earth. In some places, such as the ridge running roughly north-south in the mid-Atlantic Ocean, mantle material flows up as molten rock, joining the plates and shoving them apart. In other areas, such as the deep ocean trench east of Japan, one plate strikes another and burrows under it back into the mantle. The crustal plates thus scrape along, slam into, and slide under each other. Their boundaries exhibit the most intense volcanic and seismic activity, and most earthquakes are recorded there. The continents, riding on the plates, are thus carried about over the surface of the earth.

We can now directly measure the motions of the continents. Laser-satellite experiments show that North America and Europe are separating at about two centimeters per year. But the evidence is convincing that they were once in contact. To separate to the twenty-five-hundred-mile breach across the Atlantic at only two centimeters per year (less than one inch per year) has taken approximately 200 million years.

Magnetic Pole Reversals

We do not need to *assume* that the drifting of the continents has been at a constant rate over the past 200 million years, for we have a magnetic record of their movements. The molten rock that oozes up in the Mid-Atlantic Ridge is partly magnetized by the magnetic field of the earth, and as the rock cools that magnetism is frozen into it. We can measure the magnetic polarity of the rocks in the sea floor with delicate magnetometers towed on the ocean surface. This would, of course, not be interesting if it were not that the magnetic field of the earth reverses itself every half million or so years; in fact, in the past 76 million years it has reversed itself 171 times. Ocean floor rocks close to the Mid-

Atlantic Ridge carry the magnetism of the present field, but those several miles away on either side have the reversed magnetic polarity of the field of the earth half a million years ago, and those still farther have the polarity of the still earlier field, and so on. Over recent geological times the alternating polarity of the magnetism in the ocean floor has reversed many times in consort with the changing field of the earth. Moreover, this pattern of field reversals is consistently reproduced in the ocean floor in widely separated regions of the earth.

But how do we know the dates of the magnetic field reversals in the ocean floor? The age of *any* reversal can be estimated by extrapolation along the sea-floor magnetic strips, assuming constant spreading rates, once a reliable chronology is constructed for a portion of the strip. Such a reliable chronology has been constructed for the most recent segment —up to 5 million years ago. It is based on direct radioactive dating of sequences of lava flows.

Radioactive Dating

Many atomic nuclei are unstable and spontaneously convert to other nuclei with the emission of particles, such as electrons (beta particles), photons (gamma rays), or even the nuclei of light atoms, such as helium. For any given nucleus, this decay process, called *radioactivity,* is random, but for each kind of radioactive nucleus there is a specific time period, called the *half life,* which describes its rate of decay. A particular nucleus may last a shorter or longer time than its half life, but in a large sample of nuclei, almost exactly half will have decayed in one half life, and half of those remaining will have decayed in two half lives (three quarters in all). After three half lives, only one eighth of the original sample remains, and so on. Thus radioactive elements provide accurate nuclear clocks; by comparing the relative abundances of a remaining radioactive element and of the element it decays to, we can learn how long the process has been going on and hence arrive at the age of the sample.

Most of the nuclei heavier than uranium have very short half lives—far, far less than one second; this is why they are not found in nature, although they can be produced for brief instants in the nuclear physics laboratory. (Even the relatively stable plutonium—atomic number 94—has a half life of only about twenty-four thousand years, very short compared with the age of the earth.) On the other hand, the earth's crust contains radioactive elements that decay slowly. Among these are potassium 40, which decays to argon 40 with a half life of 1,250 million years, rubidium 87, which decays to strontium 87 with a half life of 4,880 million years, and uranium 238, which decays through a series of elements (including radium) to lead 206 with a half life of 4,470 million years. These nuclear clocks enable us to measure the ages of the rocks in which these elements are found.

There are few rocks on the earth older than 2,800 million years; the oldest yet found (in western Greenland) have ages of 3,900 million years. The ages of these rocks give a lower limit to the age of the earth itself. On the other hand, lunar rocks, returned by Apollo astronauts, and meteorites have been found with ages of up to 4,500 million years. Analysis of such evidence, along with information gleaned from our planetary probes on early accretion and melting at the surfaces of the moon and planets, lead us to believe that the bodies of the solar system, including the earth, formed during a 100-million-year period that began 4,600 million years ago.

Radiocarbon Dating

A particularly important method of radioactive dating over relatively short periods (forty to fifty thousand years) is that of radiocarbon dating, developed by W. F. Libby and his associates in 1947. The nucleus of the most common kind of carbon atom, called carbon 12, contains six positively charged protons and six electrically neutral neutrons. A tiny fraction of carbon nuclei, those of carbon 13, contain six protons but seven neutrons. Both carbon 12 and carbon 13 are stable (unradioactive). But another kind of carbon, carbon 14, has six protons and eight neutrons in each of its nuclei. Carbon 14 is radioactive, decaying to ordinary nitrogen (nitrogen 14) with a half life of about 5,730 years. With so short a half life, any carbon 14 that may have been around when the earth was formed has long since decayed.

Carbon 14, however, is continually being formed in the air from nitrogen 14 by the action of cosmic rays, atomic nuclei from space that strike the earth's atmosphere with tremendous energy. The newly formed carbon 14 immediately oxidizes to form carbon dioxide, and this carbon 14 carbon dioxide diffuses and mixes with the common carbon dioxide containing carbon 12 nuclei. An equilibrium is reached between the formation and decay of the carbon 14, so that about one molecule in one million million molecules of carbon dioxide contains carbon 14.

Now the carbon dioxide in the atmosphere is absorbed by green vegetation in the process of photosynthesis; the oxygen is released into the atmosphere, and the carbon goes into building the plants. Animals eat plants (and each other), and that same carbon helps build their bodies as well. Thus the carbon in all living animal and plant tissue has about one atom in one million million that is carbon 14. But when the plants and animals die, they stop absorbing new carbon dioxide and no new carbon 14 accumulates. That present at the time of death gradually decays, half by the end of the first 5,730 years, three quarters in 11,460 years, and so on. Because of its radioactivity, carbon 14 can be detected with radiation counters, and in a sample of given mass the intensity of the radioactivity indicates how much carbon 14 is left. Thus we can learn the ages of plant and animal remains, such as wood and mastodon bones.

Radiocarbon dating is not precise. One reason is the difficulty of measuring the tiny concentration of carbon 14. The older the sample the less carbon 14 is left, and thus the greater the uncertainty. For a sample with an age of forty thousand years, the uncertainty is typically about 5 percent. Another problem is that the influx of cosmic rays is not absolutely constant, so the production rate of carbon 14 varies slightly. The variation is extremely small, however, and, moreover, radiocarbon dating can be calibrated from the ages it gives for objects whose ages are known from other information—for example, the thirty-three hundred-year-old wooden coffin of King Tutankhamen or the bristlecone pines.

Radiocarbon dating has been especially useful in finding ages of fossils of relatively recent origin—trees, mammals, and especially recent humans. In fact, burnt wood found in the hearths of prehistoric peoples has been of great use in working out man's recent history and in relating it to times of climatic change on earth. (Dinosaurs are far too old for radiocarbon dating; they died out more than 60 million years ago, and we must deduce their ages from those of the rock layers where we find their bones.)

The fossil record is discussed elsewhere in this book, but I mention it here to emphasize that we have sound ways of establishing the ages of plant and animal remains. We find that humans (genus *Homo*) date to approximately 2 million years ago (although remains of the closely related genus *Australopithecus* are found in deposits roughly twice that age), while the earliest blue-green algae are 3.5 thousand million years old.

Of course there are uncertainties about the ages of individual specimens, up to 10 or even 20 percent. But no uncertainty could allow an error of roughly 3,000 million years! If you see an aged man with a white beard, deep wrinkles, and a bald head, you may not know whether he is seventy- or ninety-years-old, but there is no question of his being a three-day-old infant. But even that would only be an error of a factor of ten thousand; the creationists, who would have the earth created only six thousand years ago, want an error of a factor of a million! Moreover, do not think that scientists' assertions are based on a few dozen old bones. There are literally tens of millions of fossils under study in museums and laboratories throughout the world.

The Age of the Solar System

I have described how radioactive dating reveals the ages of terrestrial rocks. In addition, the six American Apollo missions and three unmanned Soviet Luna missions all brought moon rocks back to earth for direct analysis. As stated above, analyses of the ages of earth and moon rocks lead to a firm estimate of 4.5 thousand million years as the age of the earth and the moon.

Thousands of fallen meteorites have been recovered and dated. We believe that these objects are fragments of asteroids—tiny planets—most of which exist between the orbits of Mars and Jupiter. Asteroids frequently collide and break into small pieces. In any event, meteorites are certainly objects from another place in the solar system. Analysis of meteorite ages leads to a firm estimate of the ages of their parent bodies as being, like the earth and moon's, 4.5 thousand million years.

So the earth, moon, and (probably) asteroids, all solar system objects, have a common age. Today the ages of these bodies are obtained from radioactive dating. But there is an entirely different, although less precise, indication of the age of the solar system—namely, that provided by the theory of the evolution of the sun.

Like most stars, the sun derives its energy from nuclear fusion. Deep in its interior, where the temperatures range up to more than 10 million degrees absolute, hydrogen atoms are fusing into helium atoms (a process similar to that we hope to achieve, on a commercially economical level, in the fusion reactor of the future). Each second the sun converts some 600 million tons of hydrogen into helium, with about 4 million tons of matter destroyed in the process—converted into energy according to Einstein's famous equation relating mass and energy, $E = mc^2$. This nuclear fusion changes the chemical composition of the sun, and as a result the entire structure of the sun must alter. Since the 1950s a great amount of work has gone into the study of the evolution of stars —how they change their structures as they convert their internal hydrogen to helium. These changes in the structure of a star cause it to change its size, its total light output, and its surface temperature—those properties that we can observe. Thus we can verify the theory of stellar evolution by comparing its predictions with observations of the temperatures and brightnesses of the stars in different clusters at various stages of evolution (the stars in each cluster are of a common age and origin).

Once stellar evolution is understood, we can compare the present sun with what it must have been like before beginning its fusion of hydrogen to helium. We find, from the differences, that the sun must have been undergoing its gradual evolution for about 5,000 million years.

By a similar analysis we can find the ages of the oldest clusters of stars in our system of stars—the Milky Way Galaxy. We know how rapidly a star must be fusing helium from hydrogen to account for its present output of light and other energy; from that conversion rate, we know how rapidly it must be evolving and hence how long it has taken it to reach its present state of evolution. With some uncertainty, we find that the oldest star clusters (and hence the Galaxy) have ages of 10 to 16 thousand million years, at least twice the age of the sun.

Time of Light Travel from Stars

One of the most fundamental laws of nature is the finiteness and constancy of the velocity of light. Light travels with the highest possible speed of 186,000 miles per second (300,000 kilometers per second). The distance light goes in one second is equivalent to seven trips around the earth. Yet it takes light more than a second to reach us from the moon. Light is just one form of *electromagnetic radiation,* which also includes radio waves, infrared radiation, ultraviolet radiation, X-rays, and gamma rays; in a vacuum the same speed applies to all of these radiations.

Those of us who followed the Apollo landings on the moon were easily aware of the nearly three seconds of time required for radio waves to make the round trip to the moon. After a question from Mission Control in Houston there would be an obvious delay—more than the astronaut's reaction time—before hearing his answer; we had to wait for our message to reach him and then for his reply to return to earth. We also can experience this kind of delay in transatlantic telephone conversations, which usually are carried via earth satellite more than twenty thousand miles above the earth's surface. The round-trip travel time in this case causes only a quarter-second delay, but we can notice it if we try to talk back and forth in rapid succession.

Light takes eight minutes to come from the sun and anywhere from minutes to hours to reach us from the planets. Voyager's signals from Saturn, for example, arrived at earth more than an hour and a half after they were transmitted. We commonly measure astronomical distances by the time it takes light to traverse them. The sun is thus *eight light minutes away.* The nearest star beyond the sun (Alpha Centauri) is *four light years away;* a light year works out to nearly 6 million million miles.

To explain in detail how we calculate the distances between earth and remote stars is beyond the scope of this brief chapter. In essence, though, we use nothing more than triangulation, or surveying. The principle is like that of depth perception. When we look at a not-too-distant object, its direction as seen from one eye is slightly different from that as seen from the other eye. The closer the object, the more pronounced the effect. (Try looking at your finger first with one eye and then with the other and see it jump back and forth in direction. Try the same thing with your finger at different distances from your nose.) The brain calculates how far away the object (say, your finger) must be to account for its different directions as seen from the separate eyes. If the object is too far away, however, (say, one hundred feet) the effect is too slight and depth perception fails.

Naturally we need a larger baseline than the distance between our eyes to survey the stars. We use, in fact, the entire diameter of the earth's orbit. As seen telescopically from two vantage points nearly 200 million miles apart, the nearer stars show barely discernible shifts in direction. Careful measures have revealed the distances to thousands of stars in this way. Like depth perception, however, the technique fails for the overwhelming majority of stars, which are too far away to show shifts in direction even when observed from opposite sides of the earth's orbit. For them we use indirect methods.

Most of those stars whose distances we *can* survey directly fall into well-defined classes, with the stars in each class having about the same intrinsic brightness. We can class a star by analyzing its light with the spectrograph, an instrument that spreads white light into its rainbow of colors and photographs the resulting spectrum. The process is a little like recognizing whether we are looking at a sixty-watt bulb, a one-hundred-watt bulb, or a two-hundred-watt bulb by reading the label on its end, except that the different classes of stars display a continuous range of stellar luminosity. Thus details in the spectrum of a star—certain missing wavelengths of light—are the key to what kind of star it is and hence to its intrinsic light output.

To find the distance to a star too remote to survey directly, therefore, we examine its spectrum to learn how bright it really is and then calculate the distance it must have to account for its apparent faintness. A motorist does the same thing at night when he sees a red stop light in the distance; he knows how bright stop lights really are, and if the one he observes appears very faint, he knows it is far away and that he may not need to slow down yet—it may turn green.

So what do we find when we measure the distances between the earth and the stars? The stars extend to enormous depths in space. In fact, several hundred thousand million stars (the sun is just a typical one among them) make up a gigantic wheel-shaped system we call the Milky Way Galaxy or simply the Galaxy. Within the Galaxy the stars are light years apart, but the whole system extends over a diameter of more than one hundred thousand light years. We thus see stars and star clusters, as well as glowing gas clouds among and between the stars, that are tens of thousands of light years away. Note that we see these objects today as they were *tens of thousands of years ago* when light left them to begin its long journey across space toward our telescopes. We have many reasons for knowing that their ages are up to millions of times greater yet, but their distances from earth alone attest to figures far greater than Ussher's scant six thousand years.

Unless, that is, we are to suppose that God created all of those remote stars less than six thousand years ago and at the same time created light from them already well on its way to us. Why would God do this?

But the Galaxy is only our tiny place in the universe. Far beyond its borders are other galaxies more or less like our own. The nearest is more than 150,000 light-years away, but most are millions, tens of millions, even thousands of millions of light-years off in space. The more remote a galaxy is, the smaller and fainter it appears, but we can still recognize it and make a pretty good estimate of its distance. In the nearest galaxies we recognize individual stars; their faintnesses tell us the distances of those nearby galaxies, which enables us to calibrate the scale of galaxy distances.

With a large telescope at least a thousand million galaxies are observable. Nearly all are so far away that we see them by light that left them before man walked on our planet, indeed, from most galaxies, light that left at the time of the dinosaurs and before.

The Time since the Big Bang

There are many other pieces of evidence for a great universal age, involving such esoteric objects as quasars, gravitational lenses, and radio galaxies, all of which display phenomena that imply ages of at least millions of years. But I shall conclude with one final example—the grandest concept of all.

Analysis of the spectra of stars and galaxies tells us more than what kinds of objects they are. If a source of light is receding from us, its light waves are spread out and arrive at larger than normal wavelengths (and vice versa if the source approaches us). This so-called Doppler effect—the displacement in wavelengths of a source's light—thus tells us how fast the source is moving in our line of sight, and in which direction.

After Einstein introduced his general theory of relativity, he applied it to the universe as a whole and found that the universe should not be stable. In the 1920s several other theoreticians predicted that the universe should be expanding. If the universe is uniformly expanding, the galaxies must be moving away from each other, and the farther apart they are the faster they must separate to keep the expansion uniform. This means that remote galaxies would be moving away from us at speeds that are greater in proportion to their distances from earth. (It would *not* imply that we are at the center or that there is a center at all. For example, suppose the universe were to double in scale in a certain time; then two galaxies, say, 200 million light-years apart must move apart from each other twice as fast as two other galaxies only 100 million light-years apart in order to double their separation in that given amount of time.)

We can check the prediction of the expanding universe by looking at the Doppler shifts in the light from galaxies at various distances. This matter was first examined by Mount Wilson astronomer Edwin Hubble in 1929, and then examined more precisely by Hubble and his colleague Milton Humason in 1931. As predicted, the remote galaxies are moving away from us, and their speeds are proportional to their distances. This phenomenon is now called the Hubble law and is generally interpreted as proof that the universe is expanding.

But if all galaxies are separating at high speed today, they must have been closer together in the past and the more so the farther back one goes in time. At some point in the remote past all of the matter of the universe would have to have been packed together in an extremely dense state.

The conditions of such a hypothetical early universe were studied by theoretical cosmologists in the 1940s and 1950s. They predicted that, according to the best known laws of physics, at that early stage the universe must have been very dense, very hot, and very bright—like the inside of a star. In this hot, dense universe about a quarter of the mass would be fused to helium and almost all of the remainder would be hydrogen. Moreover, it was predicted that as the universe expanded and cooled, the radiation its gas emitted in the hot state would no longer interact with atoms and would flow freely through space. We should still be able to see that radiation today, but because of the expansion of the universe its wavelengths would be shifted from those of visible light to those of radio waves.

The present-day universe is observed to be three quarters hydrogen and one quarter helium (by mass), save for a small contamination of those heavier elements that make up things like the earth and our bodies; these heavier atoms, we think, had their origin in nuclear fusion in early generation stars.

Moreover, in 1965 faint radio radiation having just the properties expected of that originally released from the early, hot universe was discovered coming from all directions in space. In 1978, the discoverers, Arno Penzias and Robert Wilson, of the Bell Laboratories at Holmdel, New Jersey, received the Nobel Prize in physics for their finding. The evidence is very good indeed that the universe has evolved from a hot, dense state and that evolution commenced with a primordial explosion known as the *big bang.*

Just when did the big bang occur? At present we cannot say very precisely, because it depends on how rapidly the gravitation of the universe is slowing its expansion. But roughly, the current age of the universe since the big bang is simply the time it must have taken for galaxies, moving at their present speeds, to have reached their present separations—between 10,000 and 20,000 million years. Uncertain, to be sure, but in any case several million times the age espoused by the creationists.

Conclusions

Most of the great religions of the world offer far more than a story of the creation: they provide inspiration, fellowship, and a moral or ethical code by which people can (if they work at it) live together in harmony. To be sure, most religions also have a metaphysics—answers to the grand questions people ask about our origins and about the universe at large. These metaphysical explanations, however, are usually regarded as somewhat symbolic, rather than as providing literal, detailed descriptions. Even the deeply religious John Milton, as seen in the quotation opening this chapter, evidently deemed it appropriate that man should feel free to inquire of nature herself in seeking answers to the questions of origination. After all, if all of the answers were known in advance, what need would we have for science or, for that matter, for any other scholarly form of inquiry?

Our present picture of the big bang well may be incorrect—it is a subject at the frontier. And anyway, present theory tells us nothing of what occurred *before* the big bang or of how the universe got into that state in the first place. Science has no problem with not knowing all the answers (it would be out of business if it did!). On the other hand, the great religions have no problem accommodating themselves to what science *has* revealed. I see nothing in the idea of an old universe evolving from a big bang, with a younger—but still old—earth developing life, which over billions of years evolved to man, that confronts a symbolic interpretation of the story of Genesis.

Those few cults and fundamentalist churches that want to replace scientific inquiry with dogmatic authority would serve their congregations better with a consideration of the broader values of religion and the roles their churches could play in the areas of inspiration and ethics.

I will never forget a lecture I attended many years ago by a leading Protestant clergyman who was well trained in science, especially in astronomy. His lecture was a popular account of the structure of the universe and cosmology, and it was not only accurate but inspired. In the course of his talk, he remarked how our scale of extragalactic distances has had to be revised since 1925, when the existence of galaxies first was demonstrated. He pointed out that we now realize that those remote galaxies are some five times as distant as then was thought (the distance scale is still very uncertain; the factor might be nearer ten than five). He said it should be humbling to astronomers to realize that their early measurements were off by five times!

During the question period I remarked that Bishop Ussher had determined the age of the universe at under six thousand years, while we now regard it to be at least 2 million times this figure. "If a factor of five should be humbling to astronomers," I said, "a factor of two million should be all the more humbling to theologians."

"Yes," he replied, "but theologians are *supposed* to be humble!"

SUGGESTED READINGS

Abell, George O. 1980. *Realm of the universe.* 2nd ed. Philadelphia: Saunders College Pubs.

—. 1982. *Exploration of the universe.* 4th ed. Philadelphia: Saunders College Pubs.

Encyclopedia of Science and Technology. 1982. 5th ed. New York: McGraw-Hill. See 4: 88–91 and 11: 328–55, 698–700.

Weinberg, Steven. 1977. *The first three minutes.* New York: Basic Books.

Wilson, J. Tuzo. 1972. Introducer, *Continents adrift: readings from Scientific American.* San Francisco: W. H. Freeman & Co.

4

Ghosts from the Nineteenth Century: Creationist Arguments for a Young Earth*

BY STEPHEN G. BRUSH

Introduction

The recent revival of "creationism" has raised an issue that most scientists thought was settled decades ago: the validity of the multibillion-year time scale for geological history. Indeed, by insisting that not only man but the earth and the entire universe were created in six days no more than about six thousand years ago, the creationists have adopted a position that has not been scientifically respectable for the last one hundred fifty years. Though a few professional scientists one hundred years ago continued to deny that humans have evolved from "lower" forms of life, even they had already abandoned the Mosaic chronology,

*This chapter is a revision of a longer paper that had appeared under the title "Finding the Age of the Earth: By Physics or by Faith?" in the *Journal of Geological Education* 30:34–58 (1982). Reprinted with the permission of the National Association of Geology Teachers.

i.e., the doctrine that the world was created in 4004 B.C. One has to go back to about 1830 to find a time when a significant number of geologists reckoned the age of the earth in thousands rather than millions of years (Haber 1959; Albritton 1980; Dean 1981).

According to the textbook prepared by the Creation Research Society for use in high school biology courses, "Most creationists believe that the age of the earth can be measured in thousands rather than millions or billions of years" (Moore and Slusher 1970, p. 416). However, in some of their publications the creationists are ambiguous about the extent to which their "model" depends on a short time scale. Thus in *Scientific Creationism* we read that "the creation model does not, in its basic form, *require* a short time scale. It merely assumes a period of special creation sometime in the past, without necessarily stating when that was" (Morris 1974). Nevertheless, "it is true that it [the creation model] does fit more naturally in a short chronology. Assuming the Creator had a purpose in His creation, and that purpose centered primarily in man, it does seem more appropriate that He would not waste aeons of time in essentially meaningless caretaking of an incomplete stage or stages of His intended creative work" (1974, p. 136). It is interesting to note that this statement appears in the "public school edition," which claims to treat the subject "solely on a scientific basis, with no references to the Bible or to religious doctrine" (1974, p. iv).

The "theological necessity of a young universe" is explained in an article by creationist T. Robert Ingram: "To suppose a Creation untold ages ago is really to dismiss the notion of Creation, as a serious matter; and to do that, in turn, is to play down, and eventually ignore or deny, the difference between the Creator and His Creation" (1975, p. 32). Note that this "necessity" is not perceived by most religions (Wonderly 1977; a discussion among creationists on this topic is reported by Lubenow 1978). Although creationists seem to have some disagreement among themselves, most creationist materials designed for public school science class use do insist on the validity of the short time scale. Indeed, "a relatively recent inception of the earth" has been included in the definition of "creation science" in various "balanced treatment" bills, including the version that passed (and was later overturned) in Arkansas.

After devoting about ten pages to a critique of radiometric dating methods, *Scientific Creationism* presents a review of "evidence for a young earth." The most definite statement here is that "10,000 years seems to be an outside limit for the age of the earth, based on the present decay of its magnetic field" (Morris 1974, p. 158). A more recent summary of the creation model states: "The age of the earth appears to be about 10,000 years. . . . Extrapolating the observed rate of apparently exponential decay of the earth's magnetic field, the age of the earth or life seemingly could not exceed 20,000 years" (Gish and Bliss 1981, p. iii). Another publication, *The Scientific Case for Creation,* asserts that "the decay of the earth's magnetic field, of all processes, probably most nearly satisfies the necessary uniformitarianism assumptions and so probably yields the best physical estimate of the earth's age," namely, "the earth almost certainly was created less than 10,000 years ago" (Morris 1977, p. 79). Each of these publications also mentions estimates by other methods yielding hundreds of thousands or even millions of years, not, however, as if these estimates should be accepted but simply as objections to the billion-year time scale.

Despite this ambivalence, which seems to be mainly a smoke screen generated to throw doubt on the evolutionist theory without committing the creationists to a definite alternative, the director of the Institute for Creation Research has made at least one unequivocal statement on the subject. In his book *The Remarkable Birth of Planet Earth,* Henry Morris wrote: "The only way we can determine the true age of the earth is for God to tell us what it is. And since He *has* told us, very plainly, in the Holy Scriptures that it is several thousand years in age, and no more, that ought to settle all basic questions of terrestrial chronology" (1972, p. 94). (I have not found a creationist who can point out such a statement in the Bible, other than Bishop Ussher's seventeenth-century addition to the King James version.)

If the creationists only wanted to attack evolution, they could have saved themselves much trouble and embarrassment by ignoring the time-scale problem. After all, it was quite common for pious scientists in the nineteenth century to assert that an indefinite period of time intervened between the first and second days of Creation or that each day corresponded to one thousand years or to an indefinite geological epoch. In this way they could accept the results of geology (and astronomy) that indicated a long time scale without letting this affect their belief in the Special Creation of Man. But the modern-day creationists have made it quite clear that they reject any such compromise with what they consider a literal interpretation of *Genesis,* and therefore they feel compelled to argue for a "young earth" in many of their publications. Since they have chosen to do so, we will assume, in comparing the "creation model" with the "evolution model," that the former entails creation of the earth and the rest of the universe no more than ten thousand years ago, while the latter entails formation of the earth several billion years ago and a universe that is at least 10 billion years old. (To be more specific, evidence of life on earth 3.5 billion years ago has recently been reported; the earth is probably about 4.5 or 4.6 billion years old; and the "big bang" may have occurred 10 to 20 billion years ago.) Since the difference between the time scales of the two models is *six orders of magnitude*—i.e., the evolution model assumes that the world is about a million times as old as the creation model does—we don't have to pin down either of them very precisely. It seems fair to say that any good evidence for an earth older than a million years would be extremely damaging to the creation model, while any good evidence for an earth younger than 100 million years would be extremely damaging to the evolution model. (Historically, just such evidence was considered damaging to Darwin's theory when it was brought forth by Lord Kelvin in the 1860s, but it was based on assumptions later found to be wrong.)

Aside from the age of the earth itself, radioactive dating methods support evolution by showing that, in general, simple forms of life were present on earth before the more complex forms. Radioactive dating methods provide independent support for biostratigraphic methods that have been used to arrange fossils into a sequence. Such a sequence of fossils provides the scientist with one line of evidence for evolution (see Schafersman, this volume). All methods of determining age are not equally reliable; the same is true of different radioactive dating techniques. Radiocarbon dating applies mainly to objects formed more recently than fifty thousand years ago. It depends on additional assumptions about the earth's atmosphere and is considerably less reliable than the other methods we will be discussing here (see Suess 1980 for a recent review of radiocarbon dating).

Times from Stellar Distances

The creationists' claim that the entire universe was created only a few thousand years ago was thoroughly refuted by the middle of the nineteenth century, in a rather simple way. In 1838 the German astronomer Friedrich Wilhelm Bessel reported the first accurate determination of the distance of a star other than the sun from the earth. By measuring the parallax (change in apparent position as seen from earth at different places in its orbit around the sun) of the star 61 Cygni, he found that its distance is approximately 10^{14} kilometers. (This is within 10 percent of the accepted modern value.) It was known that the speed of light is about 3×10^5 km/sec, so that in a year light would travel nearly 10^{13} kilometers. (More precisely, a light year is 9.46×10^{12} km.) Thus light starting from 61 Cygni would take a little more than ten years to reach us.

By this time William Herschel and other astronomers had already estimated that the distances of some nebulae must be hundreds or thousands of times greater than those of stars such as 61 Cygni. The basis of such estimates is the physical law that the *apparent* brightness of a light source decreases as the square of its distance from the observer and is proportional to its *intrinsic* brightness (how bright it would appear at a standard distance). Several nebulae could be resolved into separate stars of very low but measurable brightness; if one assumed that one of these stars had the same intrinsic brightness as 61 Cygni, then one could determine its distance.

The first scientist to perceive the implications for creationism of these astronomical discoveries was Marcel de Serres, a professor of mineralogy and geology at the University of Montpellier in France. In his book on the creation of the earth and the celestial bodies, published in 1843, Serres pointed out that some nebulae must be at least 230,000 light years away, on the basis of the results mentioned above. It is important to note that this conclusion does not depend on the assumption that all stars have the same intrinsic brightness (which is now known to be false) but only that *at least one* star in a nebula is as bright as 61 Cygni. The only way Serres's argument could fail is if 61 Cygni just happened to be millions of times brighter (intrinsically) than *every other star* in the sky, a rather unlikely circumstance. We do not have to make any assumptions about the accuracy of other methods of measuring stellar distances, or about the rate of processes such as the expansion of the universe, in order to accept Serres's conclusion, which is simply that *some stars in the sky must have existed much more than six thousand years ago,* contrary to the creationist doctrine.

Serres went further in pointing out an absurd consequence of the creationist assumption that the entire universe was created at the same time only a few thousand years ago. In that case Adam would have seen no stars in the sky (other than the sun) for ten years; after that they would have started to appear, one by one, as their light first reached the earth. (To refine this argument in the light of more recent discoveries, we might say four years rather than ten, but we would still be granting for the sake of argument the creationist assumption that no stars now visible are more than six thousand light years distant.) Throughout recorded history, according to this hypothesis, the number of stars seen in the sky would increase every year until the present multitude was visible. Such a remarkable phenomenon, if it had really taken place, could hardly have gone unnoticed, especially by the early seafaring people who relied on the stars for navigation. A nova or supernova, rather then being such a rare event that it undermined the credibility of Aristotle's cosmology in 1572, would have been almost an everyday occurrence according to the creationist model.

How do the creationists answer this argument? Their explanation is worth quoting at length and should be read by anyone who still thinks it is possible to defend a "scientific creation model" without falling back on theology:

> If the stars were made on the fourth day, and if the days of creation were literal days, then the stars must be only several thousand years old. How, then, can many of the stars be millions or billions of light-years distant since it would take correspondingly millions or billions of years for their light to reach the earth?
>
> This problem seems formidable at first, but is easily resolved when the implications of God's creative acts are understood. The very purpose of creation centered in man. Even the angels themselves were created to be 'ministering spirits, sent forth to minister for them who shall be heirs of salvation' (Hebrews 1:14). Man was not some kind of afterthought on God's part at all, but was absolutely central in all His plans.
>
> The sun, moon, and stars were formed specifically to 'be for signs, and for seasons, and for days, and years,' and 'to give light upon the earth' (Genesis 1: 14, 15). In order to accomplish these purposes, they would obviously have to be visible on earth. But this requirement is a very little thing to a Creator! Why is it less difficult to create a star

> than to create the emanations from that star? In fact, had not God created 'light' on Day One prior to His construction of 'lights' on Day Four? It is even possible that the 'light' bathing the earth on the first three days was created in space as en route from the innumerable 'light bearers' which were yet to be constituted on the fourth day.
>
> The reason such concepts appear at first strange and unbelievable is that our minds are so conditioned to think in uniformitarian terms that we cannot easily grasp the meaning of creation. Actually, real creation necessarily involves creation of 'apparent age.' Whatever is truly created—that is, called instantly into existence out of nothing—must certainly look as though it had been there prior to its creation. Thus it has an appearance of age.
>
> This factor of created maturity obviously applies in the case of Adam and Eve, as well as of the individual plants and animals. There is nothing at all unreasonable in assuming that it likewise applies to the entire created universe! In fact, in view of God's power and purposes, it is by far the most reasonable, most efficient, and most gracious way He *could* have done it. [Morris 1972, p. 61–62]

Readers familiar with the history of the subject will recognize that Morris has revived one of the most notorious methods for explaining away the evidence of the antiquity of the earth—the *Omphalos* (navel) theory of Philip Gosse, published in 1857 (see Gardner 1957, p. 124–27; Haber 1959, p. 246–50). Gosse proposed that Adam had a navel because God created him to look as if he had been born in the usual way with an umbilical cord; and that God thus created the entire universe in such a way that it *appeared* to have a history of previous existence. In the same way we could of course assume that God created the universe ten seconds ago and that our own memories of previous experiences were created at the same time. The creationists may themselves accept such a religious doctrine in order to explain away the overwhelming evidence for biological evolution and the antiquity of the world, but they have no right to inject it into public schools under the guise of a "scientific" hypothesis.

Darwin vs. Kelvin

Before 1905 one of the scientific objections to Darwin's theory of evolution was that there had not been enough time for a process as slow as natural selection to produce all of the earth's species, since Lord Kelvin had estimated the age of the earth to be substantially less than 100 million years. Creationists have recently revived this objection, despite the fact that the original basis for Kelvin's estimate is now known to be completely wrong, and a short time scale for earth history can be maintained only by denying the validity of radioactive and other dating methods. To understand this situation we must first review briefly the Darwin-Kelvin controversy. (Further details may be found in the books by Burchfield [1975] and Brush [1978*a*, chap. III]; a good overview of the history of geological time estimates is given by Faul [1978].)

In the first edition (1859) of his *Origin of Species*, Charles Darwin conjectured that certain geological processes such as the gradual removal of solid material from chalk cliffs by water might have been going on for as long as 300 million years. This figure was introduced as a concrete example of the length of the time periods involved in natural history, even though the theory of natural selection contained no time parameters and thus provided no direct *biological* way to estimate how much time evolution would take. An anonymous reviewer (presumably a geologist) criticized this estimate and Darwin removed it from later editions, but the unfortunate impression was created that he was "retreating" from his original position that there had been ample time for evolution (Eiseley 1958, p. 245). He could just as well have started with an estimate of 30 million years, to which no one could have raised serious objections, and it would have made little difference to the biological aspect of evolution.

William Thomson (later known as Lord Kelvin) was at this time the best known of the younger generation of British physicists and was highly respected for his research on heat theory. He was also the only person in Great Britain who was thoroughly familiar with Fourier's mathematical theory of heat conduction. This theory allowed one to reconstruct the thermal history of a simple physical system such as a sphere the size of the earth, provided certain assumptions were accepted. First, the entire sphere must be a solid (so that no heat flows by convection); second, it must have constant thermal conductivity and heat capacity throughout; third, it must have been initially all at the same high temperature, surrounded by an infinite space at a lower temperature; fourth, no heat is generated or destroyed anywhere in the system. Then, from data on the present temperatures at points just below the surface (average vertical temperature gradient), one can compute the amount of time that has elapsed since the sphere was at a specified initial temperature. In this way, Kelvin estimated that a period of 100 to 200 million years might have been required for the earth, assumed initially to be at a uniform temperature of several thousand degrees (above the melting points of the rocks), to reach its present state (1862*a*).

Kelvin also attempted to estimate the age of the sun, assuming that its energy came from converting the gravitational attraction of its parts into heat and using data on its present rate of heat loss. He concluded that it is "most probable that the sun has not illuminated the earth for 100,000,000 years, and almost certain that he has not done so for 500,000,000 years" (1862*b*, p. 393).

Kelvin thought that his results should "sweep away the whole system of geological and biological speculation demanding an 'inconceivably' great vista of past time, or even a few thousand million years, for the history of life on the earth" (1897, p. 344). He did not reject the principle of evolution itself, and indeed the conflict between Kelvin and Darwin has been somewhat exaggerated by historians; it was primarily the randomness and lack of conscious direction in Darwin's theory of natural selection that was offensive to Kelvin. Evolution guided by Divine Wisdom—the "argument of design" as it was then called—would go much faster, he thought, and thus would not be hampered by the limited time scale imposed by heat-conduction calculations (1894, p. 89–90, 204–205).

The major public defender of evolution in nineteenth-century Britain was T. H. Huxley, known as "Darwin's bulldog." In a lecture in New York in 1876, Huxley pointed out that Kelvin's physical argument was entirely irrelevant to biology. It is up to the geologists and physicists, he declared, to decide on the age of the earth; once they have agreed among themselves, biologists will accept the decision. Biologists are interested only in whether evolution actually has taken place. "We take our time from the geologists and physicists; and it is monstrous that, having taken our time from the physical philosopher's clock, the physical philosopher should turn round upon us, and say we are too fast or too slow" (Huxley 1894, p. 134). This was a perfectly reasonable argument in view of the fact that evolutionary theory did not at that time *require* any particular time scale in the absence of direct biological evidence of the *rate* of evolutionary change. But Huxley's point was ignored by those who looked for a decisive battle between physics and evolution. And most geologists, rather than defending their earlier statements that geological processes had been slowly acting over very long periods (hundreds of millions of years or more), revised them to conform to Kelvin's estimates. In the 1890s new heat-conduction calculations by P. G. Tait and Clarence King reduced Kelvin's original limit of 100 million years to somewhere between 10 and 20 million years, putting considerable strain on the geological time scale.

At the beginning of the twentieth century the situation changed radically for a reason that neither geologists nor physicists could have anticipated—the discovery of radioactivity. Following the isolation of radium by Marie and Pierre Curie, it was announced by Pierre Curie and Albert Laborde (1903) that radium salts generate a substantial amount of heat. Himstedt pointed out (1904) that if there are widespread deposits of radium in the earth, the heat they produce must be taken into account in studies of the thermal history of the earth. Liebenow (1904) followed up this suggestion by estimating that the presence of 1/5,000 of a milligram of radium per cubic meter, distributed uniformly throughout the earth's volume, would be sufficient to compensate for the observed loss of heat by conduction through the crust. A similar suggestion was made about the same time by Rutherford and Strutt. Thus the possibility of continual generation of heat over long periods of time invalidated Kelvin's assumption that the earth is simply cooling down from an initial high-temperature state.

It was soon recognized that the relative proportions of lead, helium, radium, and uranium in rocks could be used to estimate their ages (Rutherford 1905, p. 485; 1906). Strutt (1905) obtained the estimate 2.4 billion years for some rocks; Boltwood (1907) found 2.2 billion years for one rock sample; the inspiration for both studies seems to have come from Rutherford. After some initial skepticism the validity of this method for estimating ages was generally recognized, along with the thousandfold increase in the time scale over that which had previously been accepted. (Historical accounts of the early period of radioactive dating may be found in the works of Burchfield [1975, chap. VI] and Badash [1968, 1969, 1979].)

Reviewing the four assumptions on which Kelvin based his estimate of the age of the earth, we see that his result was much too low because of the falsity of his fourth assumption, that no heat is generated or destroyed anywhere in the system. But the other three assumptions were also found to be incorrect in the twentieth century. Contrary to Kelvin's first assumption, the earth has a liquid core (see Brush 1980 for the history of this discovery), and even its mantle, which seems to be solid, transfers significant amounts of heat by convection (Hallam 1973, p. 75; Takeuchi, Uyeda, and Kanamori 1970, pp. 195, 229, 234). Contrary to his second assumption, the physical and chemical states of the interior are different from that of the crust, so one cannot take the thermal conductivity and heat capacity as constant throughout. And contrary to the third assumption, there is no longer any good reason to assume that the earth was initially a hot fluid that has been cooling to its present state; it is just as likely that it was formed by the aggregation of cold, solid particles and later warmed up to its present temperature (Chamberlin 1899; Brush 1978*b*; Wetherill 1980, 1981). Thus any attempt to estimate the age of the earth from its thermal properties alone is likely to be completely wrong.

The Process of Radioactive Decay

Several other methods for estimating geological time have been proposed. Each of them depends on the assumption that the rate of some process which we can measure at present has been constant (or changed in a known way) in the past. Thus if the process involves the accumulation of a physical quantity X (for example, the amount of salt in the oceans) and we assume that the time-rate of change of X has always been R and that the initial amount at time t_0 was X_0, then the "age" is simply the time interval $(t-t_0)$ during which the process is supposed to have been operating, and is given by

$$A = t-t_0 = \frac{X-X_0}{R}$$

If we assume there is none of the quantity to start with $(X_0 = 0)$, then

$$A = X/R.$$

Unfortunately, almost every process that has been suggested for this purpose is obviously affected by many physical, chemical, geological, and biological factors that have *not* been constant in the past. The only apparent exception is radioactive decay. The early experiments by Rutherford and others indicated that decay involved the actual transmutation of one element into another, for example radium into lead with the emission of "alpha particles" (helium nuclei). The measured rate of emission was found to be proportional to the amount of radium present, but did not change with pressure, temperature, or any other factors.

The discovery of radioactivity showed that Kelvin's basic assumption was wrong—radioactive substances in the earth's crust might generate enough heat to balance a substantial part of that lost by conduction out into space. Moreover, Rutherford and Boltwood suggested that the relative proportions of uranium, lead, and helium in rocks could be used to estimate the time that had passed since their crystallization, and by 1907 ages of more than a billion years had been found in this way.

The radioactive decay of a nucleus is a random process, described by a decay constant λ that gives the probability that a nucleus will decay in a definite time interval. In other words, the process of radioactive decay can be described by saying that a radioactive atom has a certain probability of decaying during a given interval of time. Half of the original nuclei will have decayed, on the average, in a time known as the "half life." The randomness of this process is a direct consequence of quantum theory and is now well established experimentally. Since the nucleus is a very small part of the atom and is insulated from the effects of other atoms by electrical repulsion, external conditions such as pressure and temperature have almost no effect on the decay constant, except in extreme cases such as inside stars.

A method frequently used to estimate the age of the earth relies on the ratios of lead isotopes, determined in a mass spectroscope. Two of these isotopes may be produced by alpha decay of uranium isotopes. A uranium 238 nucleus, containing 92 protons and 146 neutrons, may emit 8 alpha particles (each of which is a nucleus of helium containing 2 protons and 2 neutrons) and 6 electrons, thus becoming a lead 206 nucleus, containing 82 protons and 124 neutrons. Similarly uranium 235 decays to lead 207. Ordinary lead also includes isotope 204, which is not produced by any radioactive decay of another element, and isotope 208, some of which is produced by the decay of thorium. As far as is known, *chemical* or *geological* processes cannot change the relative abundances of these isotopes.

If we knew the "primeval abundances" of the lead isotopes, i.e., the relative amounts of isotopes 204, 206, 207, and 208 at the time the earth was formed, we could then subtract these from the present abundances to estimate the amount formed by decay of uranium and thorium. If there had been no chemical separation of lead from uranium since the time of the earth's formation, we could estimate the age of the earth by comparing the amount of radiogenic lead with the amount of uranium present in a rock.

Of course it is not valid to assume that no chemical separation of lead from uranium has occurred, since all rocks were probably molten at some time after the formation of the earth. The assumption *may* be valid for the period of time since the rocks were last crystallized ("time of mineralization"—t_m).

In 1946 Arthur Holmes and F. G. Houtermans pointed out that the age of the earth could be estimated from the relative abundances of lead 206 and 207 in rocks even though those rocks had crystallized more than a billion years after the formation of the earth. The key to their method is that there are *two* radioactive decays involved, each having a different decay constant; therefore, the end products of these decays—206 and 207—accumulate at different rates. Suppose one has several rocks that were formed at the same time (t_m) but that contain different proportions of uranium and lead. Then the so-called "isochron" or plot of the amount of radiogenic lead 207 against the amount of lead 206 should be a straight line. Isochrons can be graphed for any such rocks that were formed simultaneously. From the slope and intercept of the isochron one can obtain two relations between t_m , t_o (in this case, the age of the earth), and the primeval isotopic abundance ratio for leads 206 and 207. If one additional piece of information about these parameters can be found, then all three can be estimated.

Another way in which Holmes and Houtermans applied the isochron method was to construct isochrons for collections of rocks with different values of t_m. Their calculation also assumed that the primeval isotopic abundance ratio was similar to that of certain rocks with a very high relative abundance of the nonradiogenic isotope 204. In this way they estimated that the age of the earth is about 2.9 billion years.

In 1953 this estimate was revised when Patterson, Brown, Tilton, and Inghram measured the abundance of lead isotopes in some meteorites that contained very little uranium. (Thus almost all of their lead 206 and 207 should be nonradiogenic.) Taking this measurement as the primeval abundance, Patterson and Houtermans arrived at the figure 4.5 ± 0.3 billion years.

Although there was originally some doubt as to the validity of the assumption that the primeval abundance of lead isotopes in the earth is the same as that now found in some meteorites, subsequent research has confirmed it. While an age based on any single set of rock or meteorite data may be subject to criticism, the fact that many independent determinations give the same value makes it extremely unlikely that they could all be seriously wrong. The isochron method is self-checking in the sense that if the group of rocks selected does not yield a straight line plot, then one knows they were not formed at the same time and can reject them.

The best current estimates of the age of the earth all lie well within the limits of error of the original Patterson-Houtermans figure. It is remarkable that this figure has been stable for nearly thirty years, in view of the radical changes that have taken place in the earth sciences during that time.

Other methods for determining the ages of very old rocks have been used to check and reinforce those found by the uranium-lead method. One method uses the beta decay of rubidium 87 to strontium 87; another uses the decay of potassium 40 to argon 40 by capture of an orbital electron. (Details of this and the other types of radiometric age determination are given in Brush [1982] and Dalrymple [1982*a*].)

Criticisms of Radioactive Dating

At the present time, the 4.5-billion-year time scale is accepted by the overwhelming majority of scientists. This fact in itself is remarkable since most other aspects of geophysics and planetary science are still in a state of lively controversy, and it would be difficult to find such near-unanimous agreement on a problem like, for example, the origin of the moon or the mechanism for continental drift. Aside from the creationists, only a handful of legitimate scientists have questioned the accuracy of the radioactivity estimate of the age of the earth. One of these scientists is A. E. Mussett of the University of Liverpool, who argues that there is not sufficient reason to identify the primeval abundance of the radiogenic lead isotopes with the values determined from meteorites, as Patterson and others did. Because the complex geological history of the earth has affected the amounts of lead and uranium in the rocks now available for study, Mussett claims that we may never be able to determine the precise age of the earth (1970; see also Gale and Mussett 1973).

Nevertheless Mussett, who is the most severe critic among the scientists who have professional qualifications in this area, concluded that "the interval 5000 to 3500 million years ago almost certainly encompasses the formation of the earth as a separate entity within the solar system" (1970 p. 72). Moreover, he points out that the objections to estimates of the age of the earth do not apply to the ages of meteorites, whose formation can be assigned quite confidently to a period from 4,500 to 4,700 million years ago. Thus Mussett's criticism does the creationists no good at all; it only strengthens the evidence against their postulate that the entire universe was created less than ten thousand years ago.

We now turn to the creationist criticisms of radioactive dating, as presented in the books by Morris (1974) and Slusher (1981).

1. *Rocks are not dated through the study of their radioactive minerals such as uranium, thorium, potassium, rubidium. Rocks are dated, instead, by methods "worked out long before anyone ever heard or thought about radioactive dating"—the methods used in constructing the geological column* (Morris 1974, p. 133, italics added).

The geologic column *was* worked out long before radioactive dating was known, but stratigraphy only provided a means of assessing sequences of geologic strata—or, as the nineteenth-century American geologist James D. Dana stated, geology determines only "relative," not "absolute," lengths of geological ages (1880, p. 590; see also Schafersman, this volume). Some estimates of the earth's age in "years" were made, but these proved to be poor because they were based on the untenable assumption of constancy in deposition rates. The currently accepted dates for geological strata are based on radiometric procedures. Morris's assertion that they are not can be easily refuted by comparing nineteenth-century assessments of geological ages of strata with those of modern textbooks. One finds vague statements about millions or hundreds of millions of years in the nineteenth-century books, but geological methods alone never provided any basis for a billion-year time scale before the introduction of radioactive dating.

2. *Ages of rocks are based on the fossils they contain, but the dating of the latter depends on the theory of evolution; hence the use of a long time scale to support evolution is a circular argument* (see Morris 1974, pp. 134–36).

Schafersman's chapter in this volume gives a detailed refutation of the argument that the stratigraphic dating of fossils depends on the theory of evolution and that the construction of the geologic column is therefore circular. Even if valid, creationist criticism of the methods of biostratigraphy could have no bearing on the methods now used to obtain "absolute" ages of rocks. Since rocks can indeed be dated by radioactive methods without relying on fossils, as I have discussed above, these criticisms are not only invalid but irrelevant—most obviously so in the case of meteorites and of the moon, both of which have been dated very accurately even though they contain *no* fossils.

3. "*Uranium minerals always exist in open systems, not closed*" (Morris 1974, p. 140). *If uranium or lead enters or leaves the system, the calculated age would be wrong.*

The only evidence presented by Morris (1974) that terrestrial uranium minerals are not closed systems is a quotation from Henry Faul's *Age of Rocks, Planets, and Stars* that does indeed state that *some* uranium minerals are unreliable for age estimates (Faul 1966, p. 61). But Faul also states: "Rigorously closed systems probably do not exist in nature, but surprisingly many minerals and rocks satisfy the requirement well enough to be useful for nuclear age determination" (1966, p. 18). As mentioned earlier, the methods of estimating ages are self-checking: if some of the particular rocks selected for analysis were not closed systems, then the isotopic abundances would not fall on a straight line (the "isochron"). Our confidence in the present value of the age of the earth derives from the existence of a large body of self-consistent data, and the fact that because of their geological history some minerals are not closed systems does not invalidate the ages determined from other minerals.

4. *Lead 206 may be converted into lead 207 by free neutron capture, so their abundance ratio cannot reliably be used to estimate the age of the earth* (see Morris 1974, p. 141; and Slusher 1981, p. 34).

This argument was suggested by creationist Melvin Cook (1966, p. 53–62), on the basis of what he himself calls "circumstantial evidence." But that evidence was based on Cook's misreading of published data. In one case he assumed that no lead was found in a certain rock whereas in fact it had not been measured; in the other case Cook drew conclusions from the amount of lead 208 found in a rock without realizing that none of it was radiogenic (Brush 1982).

5. *"The uranium decay rates may well be variable"* (Morris 1974, p. 142).

We will consider one by one the creationists' suggested reasons for variation.

a. *Cosmic radiation produced by a supernova could have "reset the atomic clocks" and thereby invalidated the carbon 14, potassium-argon, and uranium-lead dating measurements. In other words, the decay constant may depend on cosmic rays and thus could have been affected by supernovae in the past* (see Morris 1974, pp. 142–43).

This objection is attributed to a column on scientific speculation by Frederick Jueneman (1972), who presents no evidence that such effects could be significant but merely says they might occur if Dudley's "neutrino sea" theory (see below) were valid. The last sentence of Jueneman's piece suggests that he doesn't really accept Dudley's conclusions himself; Morris fails to include that sentence in his quotation.

A recent paper by G. Robert Brakenridge (1981) of the University of Arizona suggests that a supernova explosion sometime between 8,400 and 11,300 years ago might indeed have affected the accuracy of the carbon 14 method—but in a direction opposite to that claimed by the creationists: with the estimated energy of 10^{49} ergs, radiocarbon dates would now appear 220 years too young; with the maximum possible energy of 10^{50} ergs, they would be 400 years too young.

Perrin proposed more than fifty years ago that radioactive decay may be due to the absorption of cosmic rays. This hypothesis was tested by L. R. Maxwell (1928), who took a polonium source 1,150 feet below the earth's surface to the bottom of a mine in New Jersey; he found that there was no change in activity within the limits of accuracy of his experiment (about 1 percent). Since most cosmic rays would be blocked at this depth, Maxwell's result is evidence against any suggestion that the rate of radioactive decay can be significantly influenced by cosmic rays.

b. *Changes in cosmic radiation produced by reversal of the earth's magnetic field could change the uranium decay rate* (see Morris 1974, p. 142).

No evidence is presented for this hypothesis, so it does not seem worthwhile to try to refute it. However, it should be noted that according to creationist Thomas Barnes's theory, which is the one primarily relied upon by other creationists to support their own ten-thousand-year time scale, there have not been any reversals of the earth's magnetic field. (See Brush, this volume, p.76 ff.)

c. *Decay rates can be changed by pressure, temperature, or chemical state* (see Slusher 1981, p. 20).

Slusher first attempted to support this claim by citing a review article by Indiana physicist G. T. Emery (1972). None of the experimental results mentioned by Emery, however, indicated a change of more than 4 percent. For the decays actually used to estimate the age of the earth there is no evidence for a change of more than 1 percent, which means that the earth is still at least 4 billion years old. In a recent letter, Emery stated that no significant changes had been reported since he wrote the review article, and, further, that "from what is known about decay-rate changes due to chemical and physical effects, there is no reason to doubt the accuracy of current radioactive dating results for the age of the earth" (1980; see also Hopke 1974).

In his 1981 critique of radiometric dating, Slusher relies heavily on a book by H. C. Dudley (1976). Dudley also cites Emery's review article along with some earlier papers, but he does not claim that any change in the decay rates of the uranium, thorium, rubidium, or potassium isotopes used to estimate the age of the earth has ever been observed.

Thus Slusher's criticism rests not on any direct experimental evidence, but only on Dudley's general theory of radioactive decay, which therefore deserves our consideration.

Dudley states that he rejects the modern physical theories that have "won almost unquestioning acceptance by the scientific community" and that he sees a need "to construct more complete conceptual *mechanistic* models of both the macrocosmos and the microcosmos, a la Rutherford and Michelson" (1976, p. 4). In particular he rejects Einstein's theory of relativity and wants to revive the classical "ether" under the modern-sounding name "neutrino sea." He also identifies this ether with the "subquantic medium" of L. de Broglie and J. P. Vigier (1963). Such a medium would allow one to reduce the apparent randomness of quantum effects such as radioactive decay to a determinism on the level of "hidden variables." As Dudley recognizes (1976, p. 24), if this subquantic ether really does exist, both relativity theory and quantum mechanics are wrong.

It is completely consistent with the philosophy of creationism to reject the randomness of natural processes (Morris 1974, pp. 15, 16, 22, 33, 59). But neither Slusher nor Morris informs the reader that his critique of radioactive dating methods involves the rejection of the two most spectacularly successful theories of modern physics—relativity and quantum mechanics. Nor do they seem to realize that the "hidden variables" postulate used by Dudley has been thoroughly disproved by recent experiments that reconfirm the extraordinary accuracy of quantum mechanics even in its most counterintuitive predictions (Clauser and Shimony 1978; d'Espagnat 1979). To attack the theory of radioactive decay by abandoning quantum mechanics seems almost suicidal; one can only suppose that the creationists know nothing about modern atomic physics (despite their "qualifications"), or that they hope no one will notice the absurdity of their position. Dudley himself rejects the conclusion that Slusher has drawn from his neutrino-sea theory. He points out that "since laboratory findings indicate that the maximum observed alteration of *any* decay rate *to date* is approximately 10 percent, usually less, the figure of 4.5 billion years for the earth's age seems to be a good ball park figure" (Dudley 1981).

Aside from their reliance on Dudley's ether theory, the creationists are prepared to throw out relativity theory in another area of science by developing an alternative theory of gravity. They hope that "consideration of the radiation of energy through gravitational waves might help to prove the youth of the universe, by setting an upper limit on the age of double stars, planetary systems, etc." (Barnes and Upham 1976, p. 197n). Another approach that allows one to obtain a young age for the universe and that involves rejecting relativity theory is to adopt the discredited hypothesis of Walther Ritz that the speed of light depends on the motion of the body emitting it (Slusher 1980).

d. *Evidence for past variations of radioactive decay constants comes from variations in the sizes of radioactive halos in minerals* (see Slusher 1981, pp. 25–26).

According to the Geiger-Nuttall law, there is a direct relation between the decay constant and the average distance travelled by the particle before it is absorbed by matter. If a speck of radioactive substance is enclosed in a crystal, the alpha particles emitted by the substance will be absorbed at a particular distance from the speck, producing a "halo" there due to change in the crystal structure. Variations in the radii of the halos indicate that the decay constant has changed.

The claim that variation in the radii of radioactive halos indicates a relatively young age for the earth was originally made by J. Joly in the 1920s, but was abandoned as more and more was learned about the radioactive decay series. The decay from uranium to lead involves the successive emission of alpha particles with different decay constants; therefore, several different halos should be formed. Eventually Joly himself abandoned his earlier argument and conceded that the age of the earth is "much more than a thousand million years" (1928, p. 239). (A thorough discussion of the early work on radioactive halos may be found in an article by Arthur Holmes [1931].)

Slusher cites only two other alleged discoveries of variation in halo size, both in creationist publications, while ignoring the large amount of research showing that halo radius is generally constant (Henderson et al., pp. 1934–39; Rankama 1954). He gives special emphasis to the work of Robert V. Gentry on polonium halos. Although Gentry appears to support the creationists' short time scale (Kazmann 1978), he has published research on radioactive halos in journals such as *Science* and *Nature* as well as in the creationist magazines. Gentry's approach appears to be considerably more objective than that of other creationists;

articles on anomalous K-Ar dates "only give strength to the applications of K-Ar dating as a geochronological method, in that they point out instances (ultrabasics and basalts erupted in the very deep ocean) where it is unsuitable to apply the K-Ar method. To say that because the K-Ar method should not be applied to such rocks, . . . it cannot be applied to *any* rocks is as logical as saying that because some plants cannot be used for human food, . . . all plants should not be used for this purpose. . . . We have continued to use K-Ar dating in our researches out here with caution, but without any hesitation as to its reliability when properly used" (1981).

Morris tells his reader that radiogenic argon 40 is indistinguishable from argon 40 "formed by unknown processes in primeval times and now dispersed around the world" (1974, p. 148), so it is impossible to use potassium-argon ratios to detect true ages. But, in fact, the proportion of atmospheric argon incorporated into the rock can be easily estimated by measuring the amount of argon 36 present and using the known isotopic composition of atmospheric argon to make this correction. Neither Morris nor Slusher even mentions this point, so we must assume that they are unfamiliar with one of the most elementary parts of the procedure they are criticizing.

7. *The existence of primordial polonium 218 halos in minerals indicates that the earth was not formed gradually over a long period of time but was created in a few hours "by Fiat nearly 6 millenia ago"* (see Gentry 1979).

According to Gentry, the halos he has observed in certain minerals were produced by the decay of primordial polonium 218, an isotope with a half life of only three minutes. If his interpretation were correct, it would imply that the earth was created in a few minutes, but Gentry presents no basis for a quantitative estimate of when this event occurred. While he has attempted to cast doubt on the long time scale based on radioactive dating, I have not found in any of his publications a criticism specific enough to call for a reply. There are alternative explanations for the halos that he attributes to primordial polonium (York 1979). In particular, Hashemi-Nezhad et al. (1979) showed experimentally that the diffusion of lead in mica can be rapid enough to explain the anomalous polonium halos. According to one of the experimenters in this group, "the haloes are inconsistent with creation less than tens to hundreds of millions of years ago unless one invents two easily observable but unobserved lead isomers of quite improbable characteristics" (Fremlin 1981).

Gentry does not claim that his results lead directly to a specific age for the earth, but he argues that the ratios of uranium 238 to lead 206 found in coalified wood from the Colorado Plateau could be explained by an infiltration of uranium a few thousand years ago (Gentry et al. 1976*b*, and telephone conversation, 16 September 1981). To accept his view that the infiltration event was associated with the creation of the earth would require discarding theories based on a large amount of data from many areas of science in order to explain a single isolated type of observation. It does not seem sensible to throw out well-established principles of science without having alternative principles that could explain at least most of the same observations, and no such alternative exists (Damon 1979; York 1979). This is a good illustration of the fact that no scientific theory can or must explain *all* observations and that a theory that is able to give satisfactory explanation of most observations will not be replaced unless a better one is available. Gentry's postulate of recent creation of the earth is contradicted by so many other facts that it has gained no support from other scientists who are familiar with this field. (See Dalrymple 1982*a* for further details on problems in Gentry and other creationists' critiques of radioactive dating.)

The Creationist Estimate of the Age of the Earth

Morris (1977) lists seventy estimates of the age of the earth, ranging from 100 years to 500 million years. As he points out, all are based on the "uniformitarian" assumption that a process has gone on at the same rate in the past as we observe it now. Therefore, he writes, not one of the estimates is reliable. The processes he refers to include the influx of various elements to the ocean, the cooling of the earth by heat efflux, and the erosion of sediments from continents. What he fails to note is that scientists now have definite reasons to suppose that these rates *did* vary in the past and also have definite reasons to suppose that the rate of radioactive decay *did not* vary. In 1936 Arthur Holmes carefully explained why measures of the influx of elements to the ocean (or the like) could not give accurate estimates of the age of the earth, but some creationists continue to use them in their attempt to shake the credibility of all standard dating techniques. Wonderly (1977) has shown that the best nonradiometric techniques clearly refute "young-earth" creationism.

The one method for estimating the earth's age that the creationists seem to consider reliable, which also happens to be almost the only one that gives an age anywhere near ten thousand years, is based on the measurement of the decay of the earth's magnetic field. Morris says this method demonstrates "that the earth almost certainly was created less than 10,000 years ago" (1977, p. 79).

Thomas G. Barnes, who proposed the magnetic field method for estimating the age of the earth, was appointed dean of the Graduate School of the Institute for Creation Research in 1981, following his retirement from the physics faculty at the University of Texas at El Paso. He was chairman of the committee responsible for the development of the Creation Research Society's biology textbook (Moore and Slusher 1970).

According to Barnes, "In 1883 Sir Horace Lamb proved theoretically that the earth's magnetic field could be due to an original event (creation) from which it has been decaying ever since" (1973, p. viii). This is not a correct description of Lamb's 1883 paper, which dealt only with electric currents and did not mention geomagnetism at all; Barnes assumes that the same mathematical equations apply to magnetism.

Barnes uses data assembled by Keith L. McDonald and Robert H. Gunst (1967) on the intensity of the earth's magnetic field from 1835 to 1965, but he selects only the "dipole component" for analysis. He fits the data to an exponential curve,

$$M \text{ (magnetic moment)} = M_0 e^{-t/T}$$

where M_0 is the value of M in 1835, namely, 8.558×10^{-22} amp meter^2. The time constant T is evaluated by substituting the value of M in 1965, when $t = 130$ years later, namely 8.017×10^{-22} amp meter^2. This gives

$$\frac{130}{T} = ln \frac{8.558}{8.017} = ln\ 1.0675 = 0.0653$$

$$T = 1{,}990$$

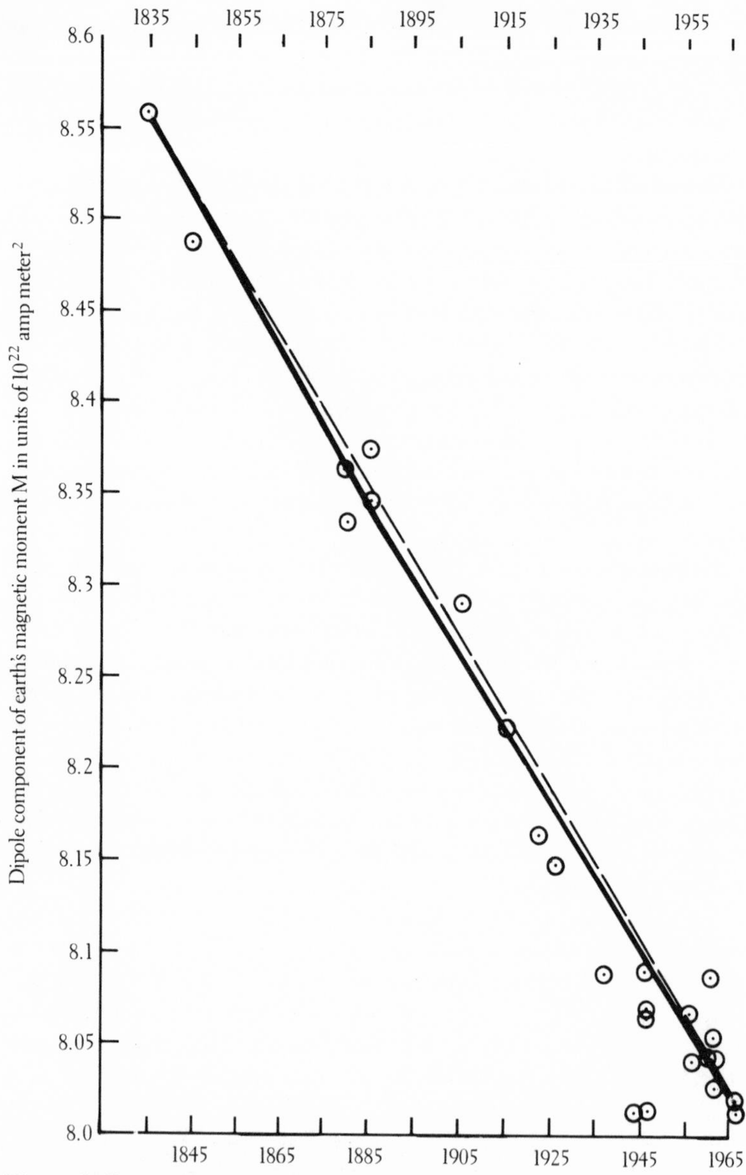

Figure 1. Observations of earth's magnetic field, 1835–1965
The dashed line is a straight line drawn between the first and last points; the solid line is an exponential curve fitted by Barnes to the first and last points. [Compiled by McDonald and Gunst (1967).]

The half life for this exponential decay would be $.693 \times 1,990 = 1,380$ years (rounded off by Barnes to 1,400 years).

Do the data actually fit this exponential formula? Barnes gives no evidence that they do; in fact, he does not even bother to present a plot showing the experimental points in relation to his theoretical curve. When one does construct such a plot (fig. 1) it becomes immediately obvious that the fit is not very good and that a straight line (shown as a dashed line) is equally good, considering the scatter of the observational points. Indeed, that is what McDonald and Gunst themselves stated: "Since the time of Gauss's measurements the earth's dipole moment has decreased, sensibly linearly, at approximately the rate of 5 percent per hundred years" (quoted by Barnes 1973, p. 34). Thus empirically there is no justification for extrapolating backwards with an exponential curve (see also Dalrymple 1982*b*).

In order to have a theoretical justification for his exponential decay curve, Barnes cannot rely on Lamb (who did not propose a physical theory of the earth's magnetism in the paper cited). He instead asserts that the earth's magnetic field is produced by currents that dissipate energy through Joule heating and that there is no energy source to restore the energy dissipated by heating.

It appears that Barnes is making the same kind of mistake Lord Kelvin made in the nineteenth century when he assumed that the earth was cooling by irreversible dissipation of energy with no source to replace the lost energy. But Kelvin made that assumption because of ignorance—radioactivity had not yet been discovered. Barnes makes a similar assumption though it is flatly contradicted both by the source of his own data, McDonald and Gunst, and by almost every geophysicist at the present time.

To begin with, McDonald and Gunst state explicitly that "the magnetic dipole field is being driven destructively to smaller values by fluid motions which transform its magnetic energy into that of the near neighboring modes rather than expend it more directly as Joule heat" (1968, p. 2057). In other words, the energy is being transferred from the dipole field to the quadrupole field and to higher moments rather than being dissipated as heat. This implies that the value of the dipole field could not have been much greater in the past, since it is limited by the total magnetic energy, which does not change very rapidly.

The other reason why Barnes's extrapolation is completely illegitimate is that we now have very good evidence for complete reversals of the earth's magnetic field in the past; over a period of thousands or millions of years, the field has been fluctuating, not decreasing. This discovery is closely associated with the development of the new theories of seafloor spreading and plate tectonics (Takeuchi, Uyeda, and Kanamori 1970; Hallam 1973; Jacobs 1975).

The theoretical basis for magnetic field reversals is Elsasser's dynamo theory, which is based on fluid motions in the earth's core (Elsasser 1946–47; see Jacobs 1975, chap. 4, or Stacey 1977, chaps. 5 and 6). The dynamo theory assumes an energy source to keep the fluid moving; it is not yet established what the main source of energy is, but there are various possibilities such as radioactive heating, growth of the inner core, differential rotation of the core and mantle, etc. In any case, nothing justifies Barnes's assumption that there is *no* energy source.

Barnes rejects the possibility that the earth's magnetic field has reversed itself, but his basis for doing so is obsolete. He cites Cowling's 1934 theorem that shows "that it is not possible for fluid motions to generate a magnetic field with axial symmetry (such as the dipole field of the earth)" (Barnes 1973, pp. 44–45). However, recent work shows that Cowling's theorem does not forbid a model with axially symmetric fluid motions generating a field with lower symmetry (Jacobs 1975, pp. 128–31), and, indeed, the earth's field does not have a pure dipole character, a fact that Barnes conveniently ignores.

The most striking example of Barnes's use of obsolete evidence is his use in a 1981 paper (p. iv) of a long quotation from a 1962 book by J. A. Jacobs on difficulties with the reversal hypothesis. In the same section of the later edition of this book, Jacobs states that "the evidence seems compelling" that such reversals *have* occurred (1975, p. 140). Barnes, however, omits the date of publication of the text he quotes from and completely ignores the fact that Jacobs changed his position in the 1975 edition. In fact, the principal creationist "expert" on geomagnetism writes as if the "revolution in the earth sciences" of the last two decades had never happened; he quotes A. A. and Howard Meyerhoff, two diehard opponents of plate tectonics, as if their "refutations" actually had been successful.

Barnes refuses to accept the overwhelming evidence for magnetic reversals not because he can find anything wrong with the evidence itself, but simply on the grounds that there is not yet a satisfactory theoretical explanation of why the reversals occur at a particular time. To support his objection he quotes from an article by Carrigan and Gubbins (1979), despite the fact that these authors express no doubt whatsoever that reversals have actually occurred.

With regard to his own theory, Barnes conveniently forgets about any need for a satisfactory theoretical explanation. He derives his short time scale by extrapolating the supposed exponential decay of the earth's magnetic field into the past, with no justification whatsoever. This leads to a value of eighteen thousand gauss for the magnetic field in 20,000 B.C., which is "stronger than the field between the pole pieces of the most powerful magnets. It is not very plausible that the core of the earth could have stayed together with the Joule heat that would have been associated with the currents producing such a strong field. [Hence] the origin of the earth's magnetic moment is much less than 20,000 years ago" (Barnes 1973, p. 38).

If any backwards extrapolation of these data were justified (which, of course, it is not), it would have to be linear rather than exponential. In order to find a field as large as eighteen thousand gauss, we would then have to make the dipole moment about sixty thousand times as great as it is now, which means we would have to go back more than 100 million years. In other words, an *empirical* analysis of the magnetic data, ignoring all theories, would immediately refute the creationists' short time scale. It is only by using a completely fallacious theory that Barnes can arrive at the conclusion that the earth is less than twenty thousand years old.

Even the linear extrapolation breaks down almost immediately. Although there are no direct measurements of the magnetic field intensity before 1835, other data can be used to estimate how it changed in the previous centuries. There is some reason to believe that the dipole field reached a maximum around 1800 and that it was smaller in 1600 than in 1800 (Yukutake 1971, p. 23). Other recent work also suggests that the dipole field has fluctuated on a fairly short time scale (Braginsky 1970; papers by J. C. Cain and others in Fisher et al. 1975). Thus there is absolutely no basis for the creationist claim that the decay of the earth's magnetic field puts a limit of ten thousand years on the age of the earth (Morris 1974, p. 158). Dalrymple 1982*b* gives a more detailed critique of the magnetic field argument.

The Age of the Moon

Finally, in a recent paper, Barnes (1982) claims that the theory of evolution can be refuted by showing that the *moon* is much less than 4 billion years old. His argument is based on the observation that the moon is now receding from the earth and therefore presumably was much closer in the distant past. If one takes the *present* rate of recession, which is based on the rate of dissipation of energy by friction in the earth's oceans, then it can be estimated that sometime between 1 and 2 billion years ago the moon would have been so close to the earth that it would have been broken up by tidal forces inside the "Roche limit." Conversely, if it had started to recede 4.5 billion years ago, it would be much farther away than it is now (Barnes 1982).

Barne's newest "disproof of evolution" sounds plausible, but it is as fallacious as his earlier arguments have been. The present rate of tidal dissipation is anomalously high because the tidal force is close to a resonance in the response function of the oceans; a more realistic calculation shows that dissipation must have been much smaller in the past and that 4.5 billion years ago the moon was well outside the Roche limit, at a distance of at least thirty-eight earth radii (Hansen 1982; see also Finch 1982). Like other creationist estimates of the age of the earth, this one is based on assuming the constancy of a process that is known *not* to be constant.

Conclusion

To support their short time scale, creationists rely on (1) the Bible; (2) estimates based on various processes such as the influx of certain elements to the ocean; and (3) decay of the earth's magnetic field. Their "biblical" evidence cannot be used in a public school science class so we ignore it. The age estimates based on natural processes are unreliable according to the creationists themselves, since we cannot assume that the rates of those processes were constant in the past. Arguments concerning the decay of the earth's magnetic field are the only ones in which the creationists' "scientific" publications seem to place much confidence.

Thomas G. Barnes developed the argument that the decay of the earth's magnetic field implies its recent creation. He assumed that the field owed its existence to electric currents that are not maintained by a dynamo or any other energy source but are decaying exponentially with time as their energy is dissipated into heat. This assumption is flatly contradicted by almost all current research on geomagnetism and is not even supported by the observational data that Barnes himself presents.

Barnes determines the parameters of his exponential decay curve from measurements of the earth's magnetic dipole in 1835 and 1965 (ignoring the fact that the decay is very nearly linear, *not* exponential, between those dates) and then extrapolates the curve backwards to 20,000 B.C. Since the field would thus have the incredibly large value of eighteen thousand gauss, he concludes that the earth must have been created after that time. But recent research in archaeomagnetism shows that the extrapolation is invalid.

To justify his calculation, Barnes has to reject the Elsasser theory (now generally accepted, at least qualitatively, by geophysicists) that the earth's magnetic field is maintained by dynamo action of the fluid motions in the core. He must also reject the conclusion (an integral part of the modern theory of plate tectonics or continental drift) that the earth's magnetic field has completely reversed itself several times in the past. The one skeptical authority on whom Barnes relies, J. A. Jacobs, has recently decided that this conclusion is now firmly established, thus undermining the basis of Barnes's calculation.

At the present time the evidence is overwhelming that the earth is several billion years old; many different determinations by radioactive dating methods give an age close to 4.5 billion years. The criticisms of radioactive dating published by creationists have no scientific basis and can be justified only by arbitrarily rejecting well-established results of modern physical science. Their argument for a young (10,000-year-old) earth from the decay of the earth's magnetic field is completely refuted by empirical data and is incompatible with all currently accepted principles of geomagnetism.

REFERENCES CITED

Albritton, Claude C., Jr. 1980. *The abyss of time: changing conceptions of the earth's antiquity after the sixteenth century.* San Francisco: Freeman, Cooper and Co.

Badash, Lawrence. 1968. Rutherford, Boltwood, and the age of the earth: the origin of radioactive dating techniques. *Proceedings of the American Philosophical Society* 112:157–69.

—. 1969. ed. *Rutherford and Boltwood: letters on radioactivity.* New Haven: Yale Univ. Press.

—. 1979. *Radioactivity in America: growth and decay of a science.* Baltimore: Johns Hopkins Univ. Press.

Barnes, Thomas G. 1973. *Origin and destiny of the earth's magnetic field.* San Diego: Creation-Life Pubs.

—. 1981. Depletion of the earth's magnetic field. ICR Impact Series, no. 100.

—. 1982. Young age for the moon and earth. ICR Impact Series, no. 110.

Barnes, T. G. and Upham, R. J., Jr. 1976. Another theory of gravitation: an alternative to Einstein's general theory of relativity. *Creation Research Society Quarterly,* 12:194–97.

Boltwood, B. B. 1907. On the ultimate disintegration products of the radioactive elements, part II. The disintegration products of uranium. *American Journal of Science* 23:77–80, 86–88.

Braginsky, S. I. 1970. Oscillation spectrum of the hydromagnetic dynamo of the earth. *Geomagnetism and Aeronomy* 10:172–81.

Brakenridge, G. Robert. 1981. Terrestrial paleoenvironmental effects of a late quarternary-age supernova. *Icarus* 46:81–93.

Broglie, Louis de and Vigier, J. P. 1963. *Theory of elementary particles.* Amsterdam/New York: Elsevier.

Brush, Stephen G. 1978*a*. *The temperature of history: phases of science and culture in the nineteenth century.* New York: Burt Franklin & Co.

—. 1978*b*. A geologist among astronomers: the rise and fall of the Chamberlin-Moulton cosmogony. *Journal for the History of Astronomy* 9:1–44, 77–104.

—. 1980. Discovery of the earth's core. *American Journal of Physics* 48:705–24.

—. 1982. Finding the age of the earth: by physics or by faith? *Journal of Geological Education* 30:34–58.

Burchfield, Joe. 1975. *Lord Kelvin and the age of the earth.* New York: Science History Pubs.

Carrigan, C. R. and Gubbins, D. 1979. The source of the earth's magnetic field. *Scientific American,* vol. 240, no. 2, pp. 118–30.

Chamberlin, T. C. 1899. On Lord Kelvin's address on the age of the earth as an abode fitted for life. *Science,* n.s., vol. 9, pp. 889–901.

Clauser, J. F. and Shimony A. 1978. Bell's theorem: experimental tests and implications. *Reports on Progress in Physics* 41:1881–927.

Cook, Melvin Alonzo. 1966. *Prehistory and earth models.* London: Max Parrish and Co.

Curie, Pierre and Laborde, Albert. 1903. Sur la chaleur dégagée spontanément par les sels de radium. *Comptes Rendus Hebdomadaires des Séances de l'Academie des Sciences, Paris* 136:673–75.

Dalrymple, G. Brent. 1982*a* Radiometric dating, geologic time, and the age of the earth: A reply to "scientific creationism." U. S. Geological Survey Publication #325. Menlo Park, Calif.

—. 1982*b*. "Can the earth be dated from decay of the magnetic field?" Menlo Park: United States Geological Survey (preprint).

Damon, Paul E. 1979. Time: measured responses. *EOS Transactions of the American Geophysical Union,* vol. 60, no. 22, p. 474.

Dana, James D. 1880. *Manual of geology.* 3rd ed. New York: Ivison, Blakeman, Taylor & Co.

Darwin, Charles. 1859. *The origin of species by means of natural selection or the preservation of favoured races in the struggle for life.* (For changes in later editions see Peckham, M. ed. 1959. *Variorum text of The Origin of Species.* Philadelphia: Univ. of Pennsylvania Press.)

Dean, Dennis R. 1981. The age of the earth controversy: beginnings to Hutton. *Annals of Science* 38:435–56.

d'Espagnat, Bernard. 1979. The quantum theory and reality. *Scientific American,* vol. 241, no. 9, pp. 158–81.

Dudley, H. C. 1976. *The morality of nuclear planning??* Glassboro, N. J.: Kronos Press.

—. 1981. Letter to Stephen G. Brush.

Eiseley, Loren. 1958. *Darwin's century.* Garden City, N. Y.: Doubleday.

Elsasser, Walter M. 1946–47. Induction effects in terrestrial magnetism. *Physical Review,* series 2, vol. 69, pp. 106–16; vol. 70, pp. 202–12.

Emery, G. T. 1972. Perturbations of nuclear decay rates. *Annual Review of Nuclear Science* 22:165–202.

—. 1980. Letter to Stephen G. Brush.

Faul, Henry. 1966. *Ages of rocks, planets, and stars.* New York: McGraw-Hill.

—. 1978. A history of geologic time. *American Scientist* 66:159–65.

Finch, D. G. 1982. The evolution of the earth-moon system. *Moon and Planets* 26: 109–14.

Fisher, R. M., Fuller, M., Schmidt, V. A., and Wasilewski, P. J. 1975. *Proceedings of the Takesi Nagata Conference, Magnetic Fields: Past and Present, June 3rd and 4th, 1974.* Greenbelt, Md.: Goddard Space Flight Center.

Fremlin, J. H. September 1, 1981. Letter to Stephen G. Brush.

Funkhouser, J. G. and Naughton, J. J. 1968. Radiogenic helium and argon in ultramafic inclusions from Hawaii. *Journal of Geophysical Research* 73:4601–4607.

Gale, N. H. and Mussett, A. E. 1970. Episodic uranium-lead models and the interpretation of variations in the isotopic composition of lead in rocks. *Reviews of Geophysics and Space Physics* 11:37–86.

Gardner, Martin. 1957. *Fads and fallacies in the name of science.* 2nd ed. New York: Dover.

Gentry, Robert V. 1973. Radioactive halos. *Annual Review of Nuclear Science* 23:347–62.

—. 1979. Time: measured responses. *EOS, Transactions of the American Geophysical Union* (29 May), p. 474.

Gentry, R. V., Cahill, T. A., Fletcher, N. R., Kaufmann, N. R., Medsker, L. R., Nelson, J. W., and Flocchini, R. G. 1976*a*. Evidence for primordial superheavy elements. *Physical Review Letters* 37:11–15.

Gentry, Robert V., Christie, Warner H., Smith, David H., Emery, J. F., Reynolds, S. A., Walker, Raymond, Cristy, S. S., and Gentry, P. A. 1976*b*. Radio halos in coalified wood: new evidence relating to the time of uranium introduction and coalification. *Science* 194:315–18.

Gish, Duane T. and Bliss, Richard B. 1981. Summary of scientific evidence for creation (parts IV–VII). ICR Impact Series, no. 96.

Haber, Francis C. 1959. *The age of the world: Moses to Darwin.* Baltimore: Johns Hopkins Univ. Press.

Hallam, A. 1973. *A revolution in the earth sciences: from continental drift to plate tectonics.* New York/Oxford: Oxford Univ. Press.

Hansen, Kirk S. 1982. Secular effects of oceanic tidal dissipation on the moon's orbit and the earth's rotation. *Reviews of Geophysics and Space Physics* 20: 457–80.

Harper, C. T. ed. 1973. *Geochronology: radiometric dating of rocks and minerals.* Stroudsberg, Pa.: Dowden, Hutchinson & Ross.

Hashemi-Nezhad, S. R., Fremlin, J. H., and Durrani, S. A. 1979. Polonium halos in mica. *Nature* 278:333–35.

Henderson, G. H. and other authors (Bateson, S., Turnbull, L. G., Mushkat, C. M., Crawford, D. P., and Sparks, F. W.). 1934–39. A quantitative study of pleochronic haloes. *Proceedings of the Royal Society of London.* A145:563–81, 582–91; A158:199–211; A173:238–49, 250–62.

Himstedt, F. 1904. Über die radioacktive Emanation der Wasser- und Ölquellen. *Physikalische Zeitschrift* 5:210–13.

Holmes, Arthur. 1931. Radioactivity and geological time. In *Physics of the Earth —IV. The Age of the Earth. (Bulletin of the National Research Council), no. 80,* pp. 124–459. Washington, D. C.: National Research Council.

—. 1936. A reply to "The earth is not old." *Report of the Committee on the Measurement of Geologic Time,* pp. 45–48. Washington, D. C.: National Research Council.

—. 1946. An estimate of the age of the earth. *Nature* 157:680–84. Reprinted in Harper 1973, pp. 124–28.

Hopke, Philip K. 1974. Extranuclear effects of nuclear decay rates. *Journal of Chemical Education* 51:517–19.

Houtermans, F. G. 1946. Die Isotopenhäufigkeiten im natürlichen Blei und das Alter des Urans. *Naturwissenschaften* 33:185–86, 219. English translation in Harper 1973, pp. 129–31.

Huxley, T. H. 1894. *Science and Hebrew tradition.* New York: Appleton.

Ingram, T. Robert. 1975. The theological necessity of a young universe. *Creation Research Society Quarterly* 12:32–33.

Jacobs, J. A. 1975. *The earth's core.* New York: Academic Press.

Joly, John. 1928. The theory of thermal cycles (A reply to Dr. Lotze). *Gerlands Beitrage zur Geophysik* 20:228–92.

Jueneman, Frederick B. 1972. Scientific speculation . . . will the real monster please stand up? *Industrial Research,* vol. 14, no. 9, pp. 15.

Kazmann, Raphael. 1978. It's about time: 4.5 billion years ago. *Geotimes,* vol. 23, no. 9, pp. 18–20.

Kelvin, Lord. 1862*a*. On the secular cooling of the earth. *Transactions of the Royal Society of Edinburgh* 23:157–70.
—. 1862*b*. On the age of the sun's heat. *Macmillan's Magazine* 5:388–93.
—. 1894. *Popular lectures and addresses.* London: Macmillan.
—. 1897. The age of the earth as an abode fitted for life. *Annual Report of the Board of Regents of the Smithsonian Institution,* July 1897, pp. 337–57 (reprinted from *Victoria Institute Transactions*).
Lamb, Horace. 1883. On electrical motions in a spherical conductor. *Philosophical Transactions of the Royal Society of London* 174:519–49.
Liebenow, C. H. 1904. Notiz über die Radiummenge der Erde. *Physikalische Zeitschrift* 5:625–26.
Lubenow, Marvin L. 1978. Does a proper interpretation of Scripture require a recent creation? ICR Impact Series, nos. 65 and 66.
L[ubkin] G. B. 1976. Mica giant halos suggest natural superheavy elements. *Physics Today,* vol. 29, no. 8, pp. 17–20.
—. 1977. Evidence for superheavies in mica looks weaker. *Physics Today,* vol. 30, no. 1, pp. 17–20.
McDonald, Keith L. and Gunst, Robert H. 1967. An analysis of the earth's magnetic field from 1835 to 1965. *ESSA Tech. Rept. IER 46-IES 1.* Washington, D. C.: Government Printing Office.
—. 1968. Recent trends in earth's magnetic field. *Journal of Geophysical Research* 73:2057–67.
Maxwell, L. R. 1928. Cosmic radiation and radioactive disintegration. *Nature* 122:997.
Moore, John N. and Slusher, Harold S. 1970. *Biology: a search for order in complexity.* Grand Rapids, Mich.: Zondervan.
Morris, Henry M. 1972. *The remarkable birth of planet earth.* Minneapolis, Minn.: Dimension Books.
—. 1974. Ed., *Scientific creationism* (public school edition). San Diego: Creation-Life Pubs.
—. 1977. *The scientific case for creation.* San Diego: Creation-Life Pubs.
Mussett, A. E. 1970. The age of the earth and meteorites. *Comments on Earth Sciences—Geophysics,* vol. 1, no. 3, pp. 65–73.
Naughton, J. J. 1981. Letter to Stephen G. Brush.
Patterson, C., Brown, H., Tilton, G., and Inghram, M. 1953. Concentration of uranium and lead and the isotopic composition of lead in meteoritic material. *Physical Review,* series 2, vol. 92:1234–35.
Rankama, K. 1954. *Isotope geology.* New York: McGraw-Hill.
Rose, H. J. and Sinclair, D. 1979. A search for high-energy alpha particles from superheavy elements. *Journal of Physics G* 5:781–96.
Rutherford, Ernest. 1905. *Radio-activity.* 2nd ed. Cambridge, Eng.: Cambridge Univ. Press.
—. 1906. *Radioactive transformations.* New York: Scribner's.
Serres, Marcel de. 1843. De la création de la terre et des corps célestes, ou examen de cette question: l'oeuvre de la création est-elle aussi complète pour l'universe qu'elle paraît l'être pour la terre? Paris: Lagny Freres.

Slusher, Harold S. 1980. *Age of the cosmos.* San Diego: Institute for Creation Research.

—. 1981. *Critique of radiometric dating.* 2nd ed. San Diego: Institute for Creation Research.

Sparks, C. J., Raman, S., Ricci, E., Gentry, R. V., and Krause, M. O. 1978. Evidence against superheavy elements in giant halo inclusions re-examined with synchrotron radiation. *Physical Review Letters* 40:507–11.

Stacey, Frank D. 1977. *Physics of the earth.* 2nd ed. New York: Wiley.

Strutt, R. J. 1905. On the radio-active minerals. *Proceedings of the Royal Society of London* A76:88–101.

Suess, H. 1980. Radiocarbon geophysics. *Endeavour* 4:113–17.

Takeuchi, H., Uyeda, S., and Kanamori H. 1970. *Debate about the earth: approach to geophysics through analysis of continental drift.* Rev. ed. San Francisco: Freeman, Cooper and Co.

Wetherill, G. W. 1980. Formation of the terrestrial planets. *Annual Review of Astronomy and Astrophysics* 18:77–113.

—. 1981. The formation of the earth from planetesimals. *Scientific American,* vol. 244, no. 6, pp. 163–74.

Wonderly, D. 1977. *God's time-records in ancient sediments: evidences of long time spans in earth's history.* Flint, Mich.: Crystal Press.

York, Derek. 1979. Polonium halos and geochronology. *EOS Transactions of the American Geophysical Union* (14 August), p. 617.

Yukutake, Takesi. 1971. Spherical harmonic analysis of the earth's magnetic field for the 17th and 18th centuries. *Journal of Geomagnetism and Geoelectricity* 23:11–31.

5

Probability and the Origin of Life

BY RUSSELL F. DOOLITTLE

Introduction

Creationists like to dote on the improbability of the spontaneous formation of the key molecules necessary for life, as well as on the problem of assembling these molecules into cells and higher organisms. They delight in quoting established scientists who have over the years pondered the problem and noted how improbable certain aspects of these phenomena seem. Unlike many of the other creationist arguments, this one is not easily parried by simply resorting to known facts. This is an area of active research where the answers are not always obvious. But there are scenarios that satisfactorily explain how these events may have taken place.

The Creationist Argument

One of the favorite creationist attacks on the concept of biological evolution concentrates on the improbability of various aspects of a spontaneous origin of life on earth. Typically, the assault begins with the timeworn watch caper. The speaker calls attention to his watch, a smoothly functioning and complicated machine with a recognized task. He allows as how, given enough study, one could figure out how the watch works. He then challenges his audience on two counts. First, he asks if anybody thinks he or she can figure out how that watch originated simply by studying its structure. He doesn't wait for an answer, though, since there might be some astute person in the audience who might venture to make a surmise along those lines. Rather, he gets right to the heart of the matter. Who out there would ever believe that this watch could have formed from a random collection of atoms? He defies his audience to bring in some iron atoms, or some silicon or carbon, or some jewels if it's an old fashioned watch, to lay the ingredients on the table, and then to wait for the materials to spontaneously arrange themselves into a watch.

Watches are not nearly as complex as living cells, the speaker continues, ignoring the point that the most marvelous aspect of life on earth is its capacity for molecular self-assembly. Instead he goes on to quote from a series of eminent scientists. He flashes a picture of a cell on the viewing screen and proclaims, "H. J. Morowitz, in his book *Energy Flow in Biology,* published in 1968, notes on page 7 that the probability of all the atoms in a simple bacterial cell arranging themselves in just that configuration at equilibrium is less than one in ten raised to the tenth power raised to the eleventh power. Why that's less than one chance in one followed by one hundred billion zeros." He repeats the last few phrases for emphasis and then continues: "Julian Huxley, in his book *Evolution in Action* notes that the chances of assembling a horse from a single cell that is allowed to mutate randomly are less than one in ten followed by three million zeroes."* At this point the creationist will usually add that Huxley's point is immaterial since it would have been impossible to form any cells in the first place.

But our man is just warming up to his task. If playing before a reasonably sophisticated audience that might contain university students or biochemists or physicians, he turns to the real persuader. He

*The reference is evidently to H. J. Muller, who is cited by Huxley (1957, 1958) as the one who calculated the absurd improbability of generating an advanced organism such as man, mammal or fruit fly fortuitously *without the operation of natural selection.* Since natural selection operates to save beneficial mutationsas they appear, Muller's calculation has no bearing on the actual evolution of man, mammal, or fruit fly [author].

will "prove" that not even simple DNA or protein molecules could come about randomly. And without those, there couldn't be life as we know it on earth, so any protests by evolutionist Huxley that natural selection solves the problem and that horses do not emerge as the result of random encounters of cells experiencing random mutations are shunted aside. Moreover, there can't be any natural selection until there is reproduction, he erroneously claims, and there can't be any reproduction until there are cells.

At this point he carries the argument to a really sophisticated stage and quotes from an article in the respected *Journal of Theoretical Biology:* "And thus we see that the commonly expressed notion of life originating spontaneously on Earth is mainly wishful thinking." H. P. Yockey, the author of the article quoted, used information theory as well as conventional probability measures in an assessment to see whether or not protein sequences of the kind observed in organisms today could ever have come about by random processes. And he concludes that they could not. What is his thesis, and how did he come to this conclusion? Yockey (1977) examined the amino acid sequence of cytochrome *c,* a protein about one hundred amino acid residues long that is ubiquitous among eukaryotic organisms, and estimated the likelihood of its evolution from random assemblages of amino acids. In essence, his approach to the problem assumes certain concentrations of amino acids in some ancient sea, using reasonable numbers that appear in the scientific literature. He then asks what the probability would be for any particular sequence appearing in a chain of amino acids brought about by random means. He adds a fillip to the calculation, which has appeared in the literature on other occasions, by suggesting that calculations based on twenty amino acids—the number found in contemporary proteins—will underestimate significantly the true probability, since nineteen of the twenty exist in two forms: mirror image structures called D and L isomers. He submits that both isomeric forms of any amino acid ought to have equal probabilities of reacting under random circumstances. This point could be debated in its own right, but it is not the major flaw in his logic.

Yockey's numbers turn out to be the following: there are approximately 3×10^{154} possible amino acid sequences one hundred residues long when thirty-nine amino acids are involved (i.e., using both D and L isomers of the nineteen amino acids that exist in two forms, plus the twentieth). Of these, he claims, only about 7×10^{60}, even allowing for the large number of amino acid replacements already observed among the cytochromes *c* of various contemporary species, will have structures that are functionally equivalent to a modern mammalian cytochrome *c*. As a result, the probability of a cytochrome *c* evolving from random origins is only about 2×10^{-94}, a value much too low to expect an occurrence, even in a billion years, anyplace in the entire universe.

As though anticipating an argument, Yockey concedes that modern proteins might have been "derived by evolution from smaller and more primitive ones." Again employing the stratagem of thirty-nine amino acids, he sets up another errant model and emerges with an estimate that the longest possible protein that could be coded for by a genome randomly evolved over the course of a billion years would only have forty-nine amino acid residues, only half the number in cytochrome *c*.

The Flaws in Yockey's Argument

The *Journal of Theoretical Biology* is refereed, but even well-reviewed journals slip up. In this case the basic premises are wrong from the start. Although Yockey's article is shot through with errors, I will limit myself to commenting on a few of the most critical flaws. First, protein evolution is not directly concerned with structures per se. It is concerned with successful function. Life as we know it does not depend on having a cytochrome *c* molecule with a particular amino acid sequence. Rather, it depends, in this instance, on having an agent that can reversibly shuttle electrons between designated agents. To this end a protein has evolved that can bind to a ring-shaped molecule called a porphyrin via two sulfur-containing amino acids and then can orient the porphyrin in an appropriate environment such that a proper approach by molecules that are electron donors or acceptors is facilitated. As I will stress when I discuss proteins, there must be an enormous number of amino acid sequences that can accomplish those ends, far higher than the estimate arrived at by Yockey with his "equivalent information content."

As for Yockey's second contention, it must be emphasized that the first peptide catalysts on earth were indeed small and relatively ineffective by modern standards. But Yockey's calculations still are completely without merit. He starts by assuming the existence of a three-letter code and, presumably, machinery for translating polynucleotide sequences into polypeptide sequences. As we shall see, these developments must have occurred well after the stage of evolution that depended most on the random assemblage of amino acids. In particular, the translation apparatus itself centers around tRNA synthetases, a family of proteins that are plainly the products of a series of duplicative events (see below). These enzymes are the ones that interface the information in polynucleotides with the product polypeptides. From the start of their existence, they probably bound only L-amino acids. D- and L-amino acids are different structures in space; mirror images do not have equal probabilities of binding to surfaces. As a consequence, the calculation that results in a maximum of forty-nine specified amino acid residues in a protein is way off, since with only twenty amino acids, instead of thirty-nine, the expected value according to Yockey's formula ought to be greatly increased.

But the article contains worse errors. The value of forty-nine was arrived at by assuming a mutation rate of 9×10^{-9} amino acid replacements per year. This value was calculated by Kimura and Ohta (1974) for modern proteins. It is based on the number that *survive* natural selection, not on the rate of occurrence. The real values for modern systems are orders of magnitude higher (Doolittle 1979). Moreover, modern organisms have exquisite editing devices for repairing errors in base-pairing during the replication of DNA and thereby keeping the mutation rate low. Clearly these could not have been operating in primitive systems. As a result, the mutation rate used in Yockey's calculation is off by at least several orders of magnitude. Thus, his maximum of forty-nine residues is totally without foundation. Ask the wrong questions, make the wrong assumptions, and you'll always proceed to wrong conclusions, no matter how sophisticated your arithmetic.

Some Scenarios for the Origin of Life

At this point let us consider some of the very real probabilistic problems that confront anyone trying to understand the origin of life on earth or other planets. Almost all investigators in the field concede that the spontaneous manufacture of small molecules—amino acids, nitrogen bases, sugars, etc.—is not a major problem in that they certainly would have been formed under a wide variety of primitive earth conditions (Miller and Orgel 1974). Putting these small molecules together into larger units does present a formidable problem, however, and considerable attention has already been devoted to its study. There have been a number of laboratory successes assembling as many as twelve nucleotides into strings under prebiotic conditions, especially when appropriate metal ions are present as catalysts (Fakhrai et al. 1981), but no one regards the problem as completely solved. Similarly, polypeptides can be formed from simple amino acids under certain conditions, but this is another area where there is much to learn. But the most puzzling problem remains the invention of coding whereby the information in a polynucleotide sequence is transferred into a corresponding polypeptide. Having taken note of these real problems, let me outline a simple scheme of the sort that may have gotten the system under way.

The first critical step is the formation of a few short strings of nucleotides, perhaps under the catalytic influence of certain metal ions. Polynucleotides have an intrinsic ability to associate with complementary strands, the nature of which I will describe below, or, if such strands are not available, to promote their manufacture. These remarkable attributes were revealed by the discoveries of Watson and Crick in the early 1950s; they are genuinely the basis of life as we know it. When Watson and Crick began their studies, it was known that nucleic acids (e.g., DNA) were composed of four different kinds of units called nucleotides (designated A, G, C and T in figure 1). These units can be combined into polymers called oligo- (meaning "some") or poly- (meaning "many") nucleotides. What Watson and Crick found was that the chemical structure of these chains is such that they wind around each other in pairs to form double spirals (double helices). They are held together by a zipper of weak bonds in such a way that the nucleotide denoted A is always opposite one denoted T; in the only other kind of match, G is always opposite C. These chemical constraints result in a quite stable structure, the two strands of which are complementary. The complementarity aspect also presents an opportunity for self-replication. As a consequence, once a few oligonucleotides had formed in some warm little primeval pond, self-assembly properties would come into play. First, an oligonucleotide serves as a template for the formation of its complement. These double-stranded forms, which tend to be stiff rods, are much more stable than the single-stranded oligonucleotides, so back reactions to the simpler monomeric units are discouraged. As it happens, however, the double-stranded forms can be unzippered in

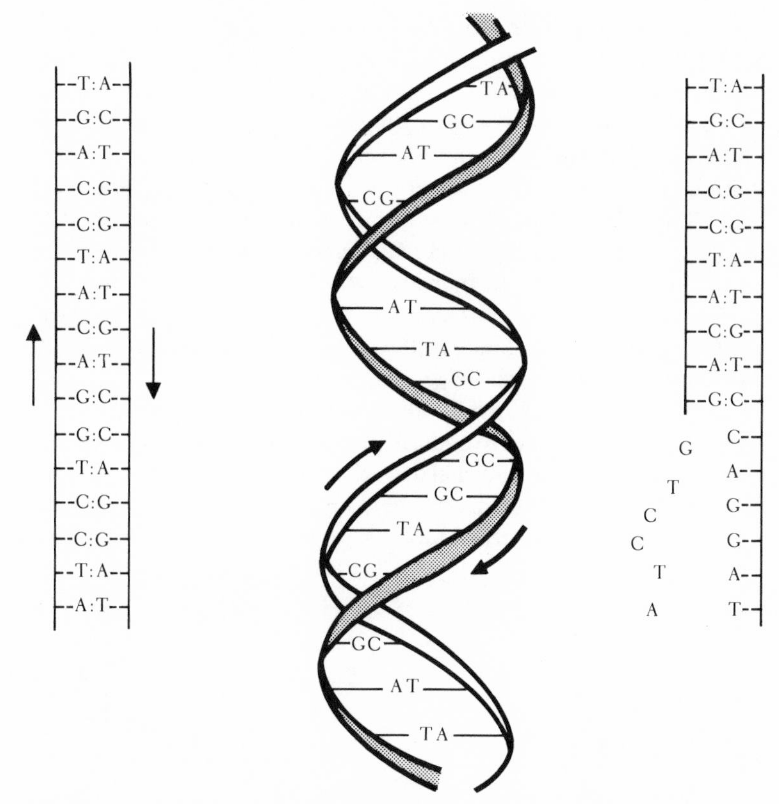

Figure 1. Schematic rendering of double-stranded DNA
In the center sketch the two complementary strands are shown in their helical modes. At left the same strands are depicted linearly. Arrows indicate that the two strands are actually aimed in opposite directions. Note that A is always opposite T and that G is always opposite C, no matter which strand has A or T or G or C. At right the formation of a new complementary strand from individual nucleotides using the preexisting strand as a template is shown.

certain ways, including by merely warming them up to 80–90° centigrade; upon cooling they reassociate. On the other hand, if they drift apart some distance during the heating period, they can serve as templates for new complementary copies. Self-replication is under way.

But even more magic remains in these simple complementary oligonucleotides. They can grow in length by the same simple processes. For example, two oligonucleotides with sequences that overlap over only portions of their structures can associate with dangling exposed single-stranded ends. These can now serve as templates for further growth (fig. 2). Thus, by these simple bootstrapping processes the nucleic acids would soon achieve substantial lengths. Marvelously also, the template-directed synthesis is not 100 percent accurate; occasionally a T is put in where a C ought to be, for example. This simple mutation process is what leads to diversity. Clearly, the complementary strands of nucleic acids are the Promethean sticks that were rubbed together to produce the spark of life.

Proteins have different but no less wonderful properties than nucleic acids. Like nucleotides, amino acids can be assembled into long chains. The character of the more diverse amino acids is such, however, that polypeptide chains fold up into unique configurations that depend only on the amino acid sequence and the properties of the solvent. The vast number of three-dimensional shapes that ensue is the secret of their catalytic power. They are able to bind other molecules, or themselves in the case of many structural proteins, very specifically because of the

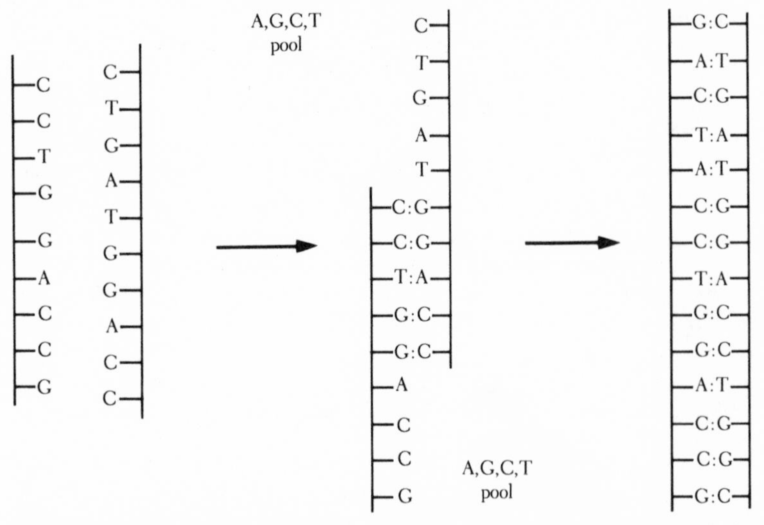

Figure 2. BOOTSTRAPPING
A simple scheme showing how two polynucleotide strands that are complementary over only parts of their structure can still pair up and lead to longer polynucleotides. The process is called "bootstrapping."

unique geometries that ensue upon folding. Instead of thinking of proteins digitally in terms of amino acid sequences, we ought to in this case think of them as the material in a ball of plasticene that can be molded into an infinity of very precise forms. The shaping hand here is natural selection operating on amino acid substitutions.

As noted above, the early protein catalysts most likely were both small and primitive by modern standards. Their invention and evolution doubtless occurred by more than one route and had different degrees of impact on the development of life in general at different stages. In very early times, polypeptides formed by simple thermal polymerization (Fox and Krampitz 1964) may have provided a general catalytic background for the formation of certain key substances. But it wasn't until protein manufacture was coupled to the information in polynucleotide sequences that the roots were struck for genuine living systems.

This coupling is certainly the most challenging problem in understanding the origin of life. An ingenious model has been proposed for bringing it about (Crick et al. 1976), the details of which are rather complicated. The crux of the proposal, however, involves the direct recognition of different amino acids by polynucleotides of similar but different sequences. These polynucleotides have sequences such that they can form internal weak bonds and thus have unique shapes. The final result of these devisings is a crude translation scheme in which a sequence of nucleotides in a polynucleotide can be translated into a definite sequence of amino acids in a polypeptide. At this point the key is to generate polypeptides with structures that can catalyze reactions. The most important of these reactions would be the polymerization of nucleotides, since that reaction yields a self-reinforcing system:

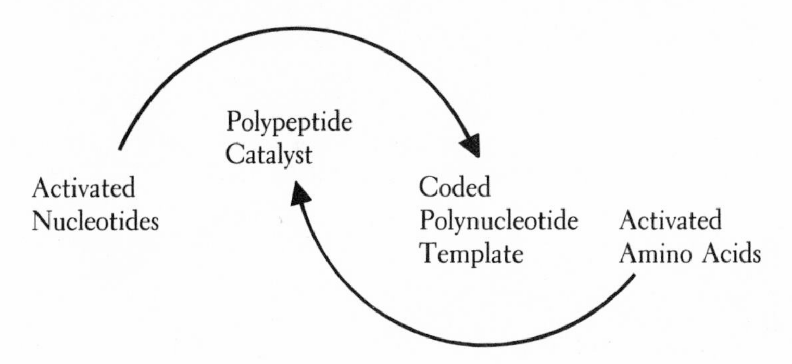

How likely is it that a population of spontaneously replicating and mutating polynucleotides would generate an appropriate sequence for the formation of a polypeptide catalyst? Rather than fall into Yockey's trap of dealing with absolute numbers for a process whose every mechanism and condition we don't yet know, let me respond with an analogy based on another set of improbable events. How likely is it that a person playing bridge will be dealt a perfect hand of all thirteen spades? The chances of this happening can easily be calculated to be one in 635,-013,559,600, the number of different sets of thirteen cards in a standard deck of fifty-two. (Of course, the probability of *any* specifically designated hand of thirteen cards is also one in 635,013,559,600). Being dealt a perfect hand in a game of bridge is obviously very improbable, but does it never happen? Of course it happens. It is merely a matter of how many people are playing over what period of time. When the game was very popular in England during the 1920s and 1930s, it is estimated that "perfect hands" were an annual occurrence in that small country alone (Weaver 1963).

And so it must have been in the molecular game of chance. We don't yet know what the appropriate "hands," or sequences, were, although there were doubtless many winning combinations. And there were certainly lots of players. Moreover, it may only have taken one winner to advance the system to the next level of organization and to an entirely new game.

Once a small number of polypeptide chains had been formed that could catalyze a few critical reactions—namely, the polymerization of nucleotides and the incorporation of amino acids—things were off and running. These early coded proteins were likely only twenty to thirty amino acids long, barely long enough to fold into a shape that could bind a nucleotide, for example. If there were three nucleotides per amino acid, as there are in contemporary systems, then the corresponding information in the polynucleotide would encompass sixty to ninety nucleotides. The catalytic power of the peptides was likely very low, but *any* increase in rate at all would drive the system forward. Once a basic set of such polynucleotide-coded catalysts existed, enzymes thereafter would be generated by polynucleotide duplications and subsequent nucleotide substitutions. This is because it is usually easier to duplicate and modify existing nucleotide-polypeptide systems than it is to make them *de novo* from random beginnings. There is an overwhelming amount of

Ferredoxin
(C. pasteuranium)

1 . . . 10 . . . 20

A Y K I ***A D*** S ***C*** V S ***C G*** A ***C A*** S E ***C P V*** N ***A*** I S ***Q*** G D . . .

. . . S I F V I D ***A D*** T ***C*** I D ***C G*** N ***C A*** N V ***C P V*** G ***A*** P V ***Q*** E

30 . . . 40 . . . 50 . . . 55

1 . . . 10 . . . 20 . . . 27

Secretin (Pig) ***H S*** D ***G T F T S*** E L ***S*** R L R ***D S*** A ***R*** L ***Q*** R L L ***Q*** G ***L*** V

Glucagon (Pig) ***H S*** Q ***G T F T S*** D Y ***S*** K Y L ***D S*** R ***R*** A ***Q*** D F V ***Q*** W ***L*** M N T

1 . . . 10 . . . 20 . . . 29

Figure 3. The amino acid sequences of some small modern proteins that show that gene duplications have given rise to lengthening in one case and to diversity in another The sequence at the top is a protein called ferredoxin (isolated from a bacterium in this case); the second half of the fifty-five-amino acid structure is clearly the result of a duplication of the first half (or vice versa) (Eck and Dayhoff 1966). The bottom two sequences are those of two small peptide hormones, secretin (Mutt et al. 1965) and glucagon (Bromer et al. 1957). Obviously the two peptides have descended from a common ancestral type. (Each letter denotes one of the twenty naturally occurring amino acids; identities are italicized.)

evidence that this is the way most modern proteins have been invented (Doolittle 1981). Moreover, the same process of polynucleotide duplication and nucleotide substitution has been the basis for the elongation of proteins, a kind of tandem duplication of the polynucleotide leading to a protein that is not quite twice as long as the original (fig. 3). The process, which stems from random breakage of the polynucleotide chains followed by a kind of mistaken pairing that yields a longer and a shorter set of partners, often occurs repeatedly. Immense proteins can result from successive doublings.

Many interesting problems concerning the origin of life remain to be explored, including the general area of bioenergetics and the encapsulation of systems by membranes. In both these areas we have reasonable scenarios that can serve as starting points for study. With this essay I have tried to show that the arguments that are raised about the improbability of the origin of life, particularly those concerning functional proteins, are often naïve and misdirected. The forerunners of today's proteins were not formed from a random collection of amino acids anymore than cells were the result of a simple aggregation of atoms. Life on earth developed in stages, each of which was built on the stabilizing, catalytic, or replicative power of the stage before it.

If I have made my point, the next time you hear creationists railing about the "impossibility" of making a particular protein, whether hemoglobin or ribonuclease or cytochrome *c*, you can smile wryly and know that they are nowhere near a consideration of the real issues. Comfort yourself also with the fact that a mere thirty years ago (before the Watson and Crick era) no one had the slightest inkling how proteins were genetically coded. Given the rapid rate of progress in our understanding of molecular biology, I have no doubt that satisfactory explanations of the problems posed here soon will be forthcoming.

REFERENCES CITED

Bromer, W. W., Sinn, L. G., and Behrens, O. K. 1957. The amino acid sequence of glucagon. V. Location of amide groups, acid degradation studies and summary of sequential evidence. *Journal of the American Chemical Society* 79:2807–10.

Crick, F. H. C., Brenner, S., Klug, A., and Pieczenik G. 1976. A speculation on the origin of protein synthesis. *Origins of Life,* vol. 7, no. 4, pp. 389–97.

Doolittle, Russell F. 1979. Protein evolution. In *The proteins,* vol. 4, 3rd ed., eds. Hans Neurath and Robert L. Hill, pp. 1–118. New York: Academic Press.

—. 1981. Similar amino acid sequences: chance or common ancestry? *Science* 214:149–59.

Eck, Richard V. and Dayhoff, Margaret O. 1966. Evolution of the structure of living relics of primitive amino acid sequences. *Science* 152:363–66.

Fakhrai, H., van Roode, J. H. G., and Orgel, L. E., 1981. Synthesis of oligoguanylates on oligocytidylate templates. *Journal of Molecular Evolution* 17:-295–302.

Fox, Sidney W. and Krampitz, Gottfried. 1964. Catalytic decomposition of glucose in aqueous solution by thermal proteinoids. *Nature* (London) 203:-1362–64.

Huxley, Julian. 1957. *Evolution in action.* New York: New American Library (Mentor).

—. 1958. The evolutionary process. In *Evolution as a process,* 2nd ed., eds. J. S. Huxley, A. C. Hardy, and E. B. Ford, pp. 1–23. London: Allen and Unwin.

Kimura, Motoo and Ohta, Tomoka. 1974. On some principles governing molecular evolution. *Proceedings of the National Academy of Sciences, USA* 71:-2848–52.

Miller, Stanley and Orgel, Leslie E. 1974. *The origins of life on the earth.* Englewood Cliffs, N. J.: Prentice-Hall.

Morowitz, Harold J. 1968. *Energy flow in biology.* New York: Academic Press.

Mutt, V., Magnusson, S., Jorpes, J. E., and Dahl, E. 1965. Structure of porcine secretin. I. Degradation with trypsin and thrombin. Sequence of the tryptic peptides. The C-terminal residue. *Biochemistry* 4:2358–62.

Watson, James D. and Crick, Francis H. C. 1953. Molecular structure of nucleic acids. *Nature* (London) 171:737–38.

Weaver, Warren. 1963. *Lady luck: the theory of probability.* Garden City, N. Y.: Doubleday.

Yockey, Hubert P. 1977. A calculation of the probability of spontaneous biogenesis by information theory. *Journal of Theoretical Biology* 67:377–98.

6

Thermodynamics and Evolution

BY JOHN W. PATTERSON

Introduction

Henry Morris, director of the Institute for Creation Research (ICR), has joined several other engineers to make thermodynamics a cornerstone of the creation-evolution controversy. For twenty years Morris has maintained that the second law of thermodynamics directly contradicts evolution. This is a simple allegation with profound apparent consequences. After all, the second law of thermodynamics was formulated well before Charles Darwin published his *Origin of Species*. Morris has argued that the century and a quarter of research in the field of evolutionary biology amounts to naught; effort could have been more wisely channeled into productive endeavors had someone only recognized the contradiction he points out. Could the second law of thermodynamics contradict evolution? Could the disciplines of thermodynamics and biology have been soisolated from one another that a paradox of such import could have gone unrecognized for over a hundred years? Is there, indeed, a paradox at all?

The answer to this question is, quite simply—no! Morris and his colleagues have constructed a completely fallacious and deceptive argument. Central to their reasoning is the notion that "uphill" processes cannot occur naturally. In making their case, they have first exaggerated the extent to which evolution is an "uphill" process (see chapter by Raup, this volume). But, granting that evolution does involve some uphill changes, it is nevertheless false that these contradict any principle derived from the study of thermodynamics. In order to demonstrate this, it will be necessary to discuss the first two laws of thermodynamics and some of the technical terms used in connection with them.

Can Water Run Uphill?

In his book *The Troubled Waters of Evolution*, Henry Morris develops an argument against evolution using a water flow analogy. His chapter "Can Water Run Uphill?" alleges that evolution to higher forms is as impossible as water pumping itself uphill, because both contradict the second law of thermodynamics (the "entropy principle"). Beneath a frontispiece photo of a picturesque waterfall, the following caption appears:

> Evolutionists have fostered the strange belief that everything is involved in a process of progress, from chaotic particles billions of years ago all the way up to complex people today. The fact is, the most certain laws of science state that the real processes of nature do not make things go uphill, but downhill. Evolution is impossible!
> [1975, p. 110]

And then just a bit "downstream" in the text we find the following passage:

> There is . . . firm evidence that evolution never could take place. *The law of increasing entropy* is an impenetrable barrier which no evolutionary mechanism yet suggested has ever been able to overcome. Evolution and entropy are opposing and mutually exclusive concepts. If the entropy principle is really a universal law, then evolution must be impossible. [1975, p. 111]

The second law of thermodynamics does indeed describe an overall tendency in nature toward decay. While the first law asserts that no energy is ever lost, the second law states that energy will always tend of its own accord to pass from a more useful or available form to a less available status. Because of this, there are countless spontaneous processes going on all the time, and they are downhill processes in just the sense that Morris and the creationists explain. However, this does not at all rule out the possibility that backward or uphill processes can take place in nature. In fact, it provides the very energy fluxes which can drive the uphill building processes that occur in life and evolution (Broda 1975; Bronowski 1970; Schrödinger 1945). The point is that a great many processes in nature are coupled to predominant downhill fluxes and are coupled in such a way that they are actually driven in the backward or uphill direction. As long as the downhill flows, off of which they feed, *exceed* the build-up or construction processes, no violation of the second law is involved. As long as this excess prevails, the net effect of the coupled processes is in the downhill direction; that is to say, they result in a net *increase* in entropy. Thus, living, evolving organisms as well as whole segments of the biosphere are viewed as "open systems" that can proceed to configurations of much lower entropy merely by feeding off the downhill fluxes that abound in nature.

The thermodynamicist uses somewhat abstruse language—the language of "classical" thermodynamics—to say this same thing. One finds such concepts as "systems," "processes," "extensive variables," and so on defined in a very technical way. Many people, unfamiliar with the jargon of thermodynamics, are easily misled by those who would distort the subject to their own advantage. To mitigate this deceit, I will provide a rather simplified description of these basic laws and concepts and convey an understanding of them in terms of familiar examples and analogies.

The Laws of Thermodynamics

The science of thermodynamics is said to have begun in the early 1800s with an analysis of heat engines by the great engineer Sadi Carnot (1824); (see Feynman, Leighton, Sands 1963). Its two basic laws, however, have since been stated and restated in several alternative but equivalent formulations. One such formulation was that of Clausius (1865), which can be translated as follows:

First law: The Energy Inventory of the World is Constant
"Die Energie der Welt ist Constant"

Second law: The Entropy Inventory of the World Tends to a Maximum
"Die Entropie der Welt strebt einem Maximum Zu"

These laws succinctly characterized what was then known about the theory of heat engines and the processes for getting work from heat. Remarkably, it has since been found that all other processes in nature conform to these laws, including the life processes.

The first law deals with the conservation of *energy,* a defined quantity that changes from one form to another. One of the most fruitful and widely applicable conservation principles of science, its logical structure is similar to the principle of conservation of *mass.* That principle was discovered years earlier, in the 1770s, by Antoine Lavoisier. Lavoisier applied meticulous accounting methods to the study of combustion, smelting, and calcination (Gale 1979) and thereby converted alchemy into the quantitative science now known as chemistry.

Unlike the first law, the second law of thermodynamics involves no conservation principles whatever. It deals only with the directionality of natural processes and uses *entropy,* another defined quantity, to determine which directional processes are possible in nature and which are not. Any process that would imply a net decrease in the entropy of the universe is deemed impossible, whereas a net increase signals a possible process. (Nothing analogous to entropy and the second law preceded their discovery; however, analogs have since turned up in communications engineering under the rubric of information theory.)

We come now to an extremely important point, and one that is not emphasized nearly enough. It turns out that energy and entropy, like Lavoisier's masses or weights, are "inventoriable commodities"—what thermodynamicists call *extensive properties.* That is, both are subject to being transferred from place to place in the universe. Moreover, transfers in these and other extensive commodities are going on all the time. Lavoisier would have had a circus with this discovery, because his highly quantitative accounting ledgers can be used to keep tabs on all this activity. Indeed, much of thermodynamic theory is little more than working out the proper accounting methods for all the "inventory" transfers and heat and work exchanges that one may wish to consider. A *process* in thermodynamics is any action that brings about any sort of inventory change or exchange. Inventory exchanges are thought of as occurring between *systems* within the universe.

Only two kinds of systems need be mentioned here. *Isolated systems,* of which the universe is considered to be the prime example, are those that exchange absolutely nothing with their surroundings. Strictly speaking, they are the only kind to which Clausius's laws apply, and one often finds "isolated systems" substituted for "the universe" or "die Welt" in alternative (but equivalent) formulations. *Open systems* are quite different. They are all the subsystems that go together to make up the universe. These can exchange heat, work, *and* substances with each other, and there are countless examples of them: a turbine pump, a jet engine, a simple flame, any organism, the entire biosphere of earth (or any segment thereof), and so on. A good understanding of these two kinds of systems is essential for discussions about the first and second laws.

Allow me to illustrate with the quaint metaphor of figure 1. Just think of the universe as a walled-in (isolated) courtyard under a blanket of freshly fallen snow. If no snow is coming down, this blanket simulates the energy inventory of an isolated system. It is isolated by the walls and does not increase in total amount. Even though this total amount of snow (energy) remains fixed, localized drifts (increases) and furrows (decreases) can be caused by some process such as the wind swirling about. Every subsection of the courtyard simulates an open system because snow can transfer either in or out. The regions that house the drift and furrow respectively correspond to two open systems that have mutually exchanged some energy. This example illustrates energy exchanges, but entropy exchanges are a little different.

If the snow has *not* stopped, but is still coming down, the amount inside the courtyard must increase continually, and this simulates the behavior of the entropy inventory in an isolated system. Again, however, there is nothing to rule out the formation of deep furrows (entropy decreases) at one or more locations in the yard. It's just that the decreases experienced in such regions will always be overcompensated by increases elsewhere.

The earth's biosphere, which is an open thermodynamic system, can be simulated by a subregion in the yard—one into which the wind has been channeled, thereby causing deep local furrowing. The creationists' position concerning entropy and the second law is tantamount to insisting that no localized furrowing can occur during a snowstorm.

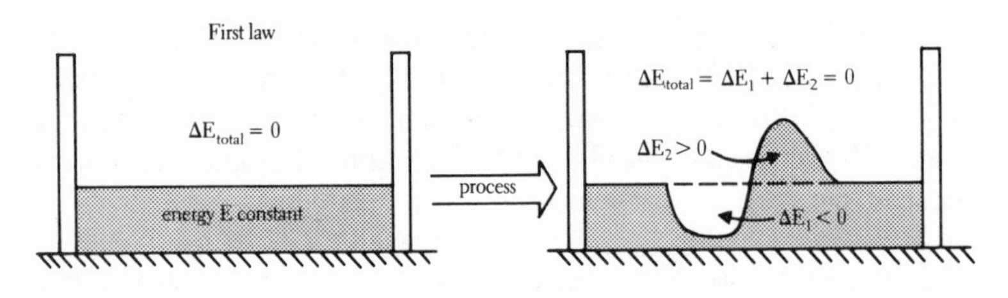

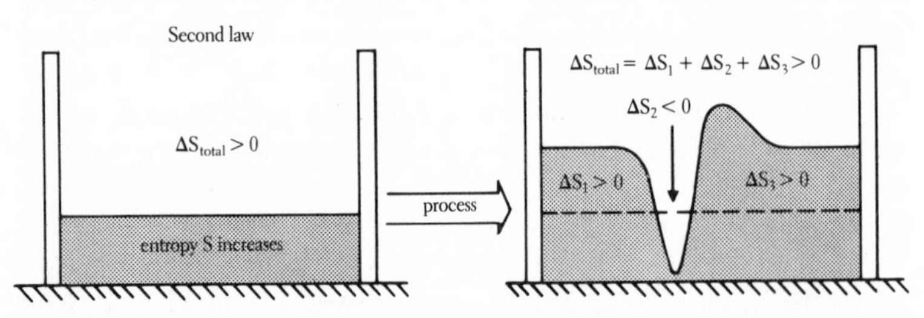

Figure 1. COURTYARD METAPHORS FOR FIRST AND SECOND LAWS OF THERMODYNAMICS
According to the first law, local energy decrease ($\Delta E_1 < 0$) may occur spontaneously in an isolated system as long as it is coupled with an increase ($\Delta E_2 > 0$) exactly compensated elsewhere; i.e., as long as the sum equals zero. It's as if the energy inventory acted like a blanket of snow inside a courtyard with the wind producing local drifts and furrows. According to the second law, the entropy decrease ($\Delta S_2 < 0$) may occur spontaneously as long as it is coupled to increases ($\Delta S_1 > 0$ and $\Delta S_3 > 0$) that overcompensate the entropy inventory nearby. Entropy inventories behave like an enclosed blanket of snow that thickens continually as snow continues to fall. However, local furrows can occur.

Localized entropy reduction is an extremely common phenomenon in living and nonliving systems alike. Indeed, it occurs each time a snowflake forms. Despite their remarkably symmetrical and highly organized structures (Bentley and Humphreys 1931), billions upon billions of snowflakes form all the time in colder climates. *Moreover, each one forms completely spontaneously and completely naturally from a completely disorganized ensemble of airborne vapor molecules!* No two designs are ever identical, nor is there any genetic code to direct a snowflake's growth or to ensure that all six of its arms will copy each other as closely as they do. We have little idea as to why the individual arms on a given flake match each other in such detail but *never* match those on other flakes (Tolansky 1958). Some day we probably will achieve better understanding of the mechanism involved, but the important point is that spontaneous complex reductions in entropy are commonplace in the natural world. Surely the creationists do not mean to argue that since the entropy principle is a universal law, snowflake formation is impossible! To be sure, scientists do not completely understand the genesis of snowflakes *or* the evolutionary process, but a declaration that either is "impossible" does not follow from the second law of thermodynamics.

Growing organisms are but additional examples of localized entropy reductions. One cannot insist that the second law of thermodynamics contradicts evolution without simultaneously maintaining that growth —development, "morphogenesis"—is similarly impossible. And evolution is merely change (via mutation, selection and other processes) in the genetic basis for morphogenesis over time. Of course, the living organism must draw its energy from its surroundings, and, of course, to maintain a highly ordered internal condition, it must rid itself of all the entropy it produces while alive. That is how the second law affects living organisms. As Levine notes in his text on physical chemistry,

> Increasing entropy means increasing disorder. Living organisms maintain a high degree of internal order. Hence one might ask whether life processes violate the second law. . . . [Indeed, there is no reason to believe they do.] The statement [that entropy cannot decrease] applies only to systems that are both closed and thermally isolated from their surroundings. Living organisms are open systems since they both take in and expel matter; further, they exchange heat with their surroundings. . . . The organism takes in foodstuffs that contain highly ordered, low-entropy polymeric molecules such as proteins and starch and excretes waste products that contain smaller, less ordered molecules. Thus, the entropy of the food intake is less than the entropy of the excretion products returned to the surroundings. . . . The organism discards matter with a greater entropy content than the matter it takes in, thereby losing entropy to the environment to compensate for the entropy produced in internal irreversible processes.
>
> [1978, pp. 123–24]

In a letter addressed to Dr. Vince D'Orazio (16 July 1979), C. R. Dawkins, senior research officer in zoology at Oxford University, expressed his dismay over Morris's assertion that the second law of thermodynamics contradicts evolution. "Morris's point about the second law of thermodynamics is pathetic," he wrote. "Surely any chemist would accept that there can be local increases in order fed by energy from outside the local system. This is what happens when you synthesize something over a bunsen burner."

How Uphill Processes Occur

It is clear that the second law of thermodynamics does not contradict the formation of snowflakes, the synthesis of chemicals over a bunsen burner, the growth and development of living organisms, or the "uphill" evolution of life in the biosphere. However, this still doesn't explain *how* uphill processes occur.

In order to do so we must leave the subject of "classical" descriptive thermodynamics, which provides descriptions of natural phenomena (laws, constraints) but does not try to explain them. As before, I will avoid rigorous discussion in favor of metaphor, in order to convey the ideas in easy-to-grasp fashion. Because engineers are so prominent in the creationist leadership, I have chosen examples that should be comprehensible to engineers, including creationist engineers.

As suggested above, an uphill or backward process can spontaneously occur in nature by being somehow coupled to a more dominant downhill process. By virtue of the coupling, the downhill process can actually drive the other process "backward" or in the so-called uphill direction. If one employs tunnel vision and focuses only on the uphill part, it may appear that any number of the basic laws are being violated, not just the second law.

Consider the example of the self-operating ram pump shown in figure 2. It was developed in Britain and put into operation during the late 1700s (Greene 1911). By constructing a very simple arrangement of conduits, self-operating flap valves and such, any competent hydraulic engineer can get low-lying water to pump part of itself *uphill* into storage tanks that may be well over a hundred feet in elevation (Graham 1943). The pumping process in this simple but primitive system will run continuously. As long as there is an adequate supply of water in the

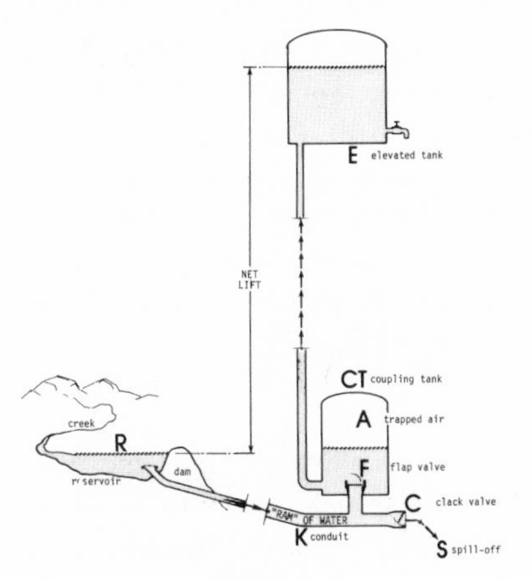

Figure 2. THE HYDRAULIC RAM PRINCIPLE
Water from the low lying reservoir R flows through the conduit K until the spring-hinged "clack" valve C snaps shut. The pressure surge thus created forces a small amount of water past flap valve F, thereby compressing the trapped air A. The pressure surge quickly terminates. Then the air A expands, keeping F closed and forcing water up the stand-pipe toward the elevated tank E. Then the spring-hinged clack valve opens again, allowing flow through conduit K to build up until the clack valve once again snaps shut. Much water is lost as spill-off S in order to elevate a very small amount to tank E. However, Graham (1943) reports that seventy-two gallons per day can be elevated over a net lift of ninety feet by a ram pump operating at 50 percent efficiency.

low-lying reservoir, it does not require any external source of power (electrical, chemical, etc.), nor are any rotary or displacement pumps used in the operation (Zebrowski 1980). All that is necessary is a gently sloping topography so that a downhill flow from the low-lying reservoir can be achieved. The conduit system taps energy from this downhill flow in order to pump a small portion to very high elevations.

Note that this rather antiquated hydraulic device lays waste to Henry Morris's (1975) analysis in "Can Water Run Uphill?" Despite his Ph. D. and his professorships in hydraulics and hydrological engineering, Morris chooses to not mention this rather widely known "hydraulic ram principle." The reason, in my view, is that the ram is spectacularly inconvenient to his argument, and, in fact, it provides a nice analog for "explaining" how natural selection works. In evolution the vast majority of mutations can be regarded as degenerative, and that same vast majority disappears. Natural selection saves the slight minority that represent an improvement. Similarly, in the ram pump, most of the water flows downhill through the pump and disappears. The flap valve in a ram pump diverts only a tiny amount to the elevated tank. In the ram, the elevation energy is derived from the downward-flowing stream. Natural selection depends upon the energy that keeps organisms alive as mutation and reproduction work their wiles. The energy requirements of living systems are derived from the foodstuffs that flow "downward" through the foodchains. Were one to neglect the downhill stream of water in a ram or the downhill flow of food energy in living systems, the attendant uphill processes would appear to violate thermodynamic laws. Taking a broader view, however, reveals that the "violations" are only superficial.

Incidentally, using Morris's argument, it might seem impossible for a sailboat to make any progress against a headwind. Such is not the case, however, as any sailing enthusiast knows. The knack of getting a proper coupling between the stern oar and the sail or "lateen" was known to sea captains as far back as the eighth century (Burke 1978).

Another excellent example of a commonplace system that exhibits uphill changes is a battery system. Under ordinary circumstances, any given battery operates in only one ("downhill") direction. If you could see inside during this mode of operation, you would see the cations (positive ions) migrating en masse toward the cathode, which becomes positive, and the anions coursing their way to the anode or negative terminal. These fluxes are set up and maintained by spontaneous chemical reactions at the two electrodes, reactions that left to themselves proceed in one way and one way only because of the energy and entropy changes that result.

Nevertheless, virtually all these unidirectional, internal processes can be made to go "backward" merely by properly coupling the battery to another one with larger terminal voltage. When this is done, we say the weaker battery is being charged. This charging is a decidedly "uphill" process. In a very real sense, it is feeding off the "downhill" processes going on in the stronger battery. Were one to observe what is going on inside a battery being charged and refuse to consider its surroundings (especially the battery charger), one might well get the impression that several basic laws of physics and chemistry were being violated—especially the law of the conservation of energy and the inflation of entropy law from thermodynamics. By taking a little larger view of the open system, however, the observer would immediately see that such is not at all the case.

If many batteries or fuel cells are joined in parallel, as in figure 3, only the strongest one of the lot operates in the discharge or downhill mode. All the others will be forced into the charging or uphill mode of operation, at least until their voltages build up to that of the charger. This serves as a useful model for multiple ion transport through cell walls and other membranes in living tissue. Such a membrane can be modeled as a system of batteries or fuel cells (one for each kind of ion) all coupled together in a common parallel bank (Patterson 1979). The ion having the greatest product of conductivity and gradient dominates the membrane as if it were the strongest of several batteries in a parallel bank. By operating in the downhill or discharge mode, this dominant-ion "battery" drives all the others backwards, at least until they achieve the null-balance, fully charged condition. This is one way that living cells

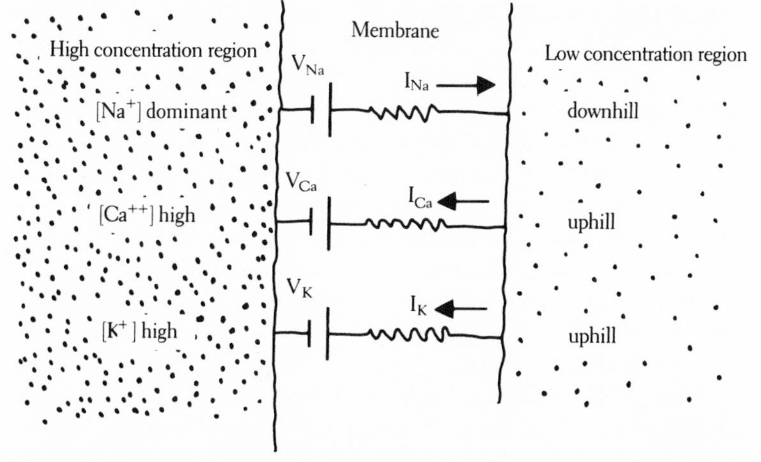

Figure 3. COUPLED CURRENT IONIC TRANSPORT THROUGH MEMBRANES
This figure illustrates multiple ion transport through membranes (modeled as a system of batteries joined in parallel). The dominant ion Na^+ (flowing to the right) flows "down" its concentration gradient, thereby setting up an electrical potential difference that drives the other mobile cations, Ca^{++} and K^+, to flow in the opposite direction—"up" their concentration gradients. (Note: V= voltage; ⊣ ⊢ =battery; I=current; ⩘⩘⩘ =resistance to current I.)

build up high internal concentrations of ions and then retain them more or less indefinitely. By now, the similarity with ram pumps should be obvious. In membranes, the dominant ion behaves like the large downhill flow that is lost from the reservoir. And, were this downhill flow not considered, the backward fluxes would appear to be violating the second law by migrating against their gradients.

The reader can improve on the above list of uphill processes that proceed by being coupled to, and hence energized by, a dominant downhill process of some sort. Coupled chemical reactions of this sort are operative in every living organism, and many similar ones have been proposed in connection with molecular evolution, the forerunner of organic evolution (Perelson 1975).

Self-organization

Closely related to the apparent "paradox" of ongoing uphill processes in nonliving systems is the apparent "paradox" of spontaneous self-organization in nature. It is one thing for an internally organized, open system to foster uphill processes by tapping downhill ones, but how did the required internal organization come about in the first place? Indeed the so-called *dissipative structures* that produce uphill processes are highly organized (low entropy) molecular ensembles, especially when compared to the dispersed arrays from which they assembled. Hence, the question of how they could originate by natural processes has proved a challenging one. As before, creationist exhortations about violations of the second law need not confuse the issue because local decreases in entropy during self-organization do not imply any such contradiction. Overcompensating increases in entropy elsewhere need only be coupled with the self-organization process. Again, the paradox is only illusory and has only to do with how self-organization occurs, not whether it does. But again we must leave the realm of classical thermodynamics to seek explanation.

Current thinking holds that self-organization can be understood in terms of theories advanced by Prigogine and his colleagues (Prigogine 1977, 1980; Nicholis and Prigogine 1977; Prigogine, Allen, Herman 1977; Glansdorff and Prigogine 1971), who have devised some very plausible explanations based upon statistical physics and new instability principles. The new instability principles apply to systems in highly nonuniform states, what Prigogine calls "far-from-equilibrium" states, but do not apply to the uniform states treated in classical thermodynamics. The new theory, for which Prigogine was awarded the 1977 Nobel Prize in chemistry, is sometimes called the theory of "non-equilibrium thermodynamics" (Lepkowski 1979; Lukas 1980) or, more often, the "theory of irreversible thermodynamics." It posits a molecular structure for matter and it inquires into the stabilities of molecular structures under various nonuniform conditions.

It has been found that the imposition of naturally occurring temperature gradients, pressure gradients, or composition gradients can force a system into highly nonuniform configurations that eventually become unstable relative to (and hence transform into) highly organized configurations. The latter would virtually *never* become stable at or near equilibrium but are actually favored when sufficiently intense gradients are imposed. Prigogine calls these highly organized configurations *dissipative structures* and a fairly wide variety have been described in theoretical analyses. Moreover, a good many have also been produced in laboratory experiments carried out in inorganic and organic media. The overwhelming majority of biochemists and molecular evolutionists who have looked into this matter realize that Prigogine's dissipative structures provide a very viable, perfectly natural mechanism for self-organization, perhaps even for the genesis of life from nonliving matter (abiogenesis). These structures can be induced merely by imposing strong temperature, pressure, or composition gradients. Indeed, those formed in certain laboratory-simulated, prebiotic broths have caused a great deal of excitement because of their remarkable similarity to the simplest known forms of life (cf., Fox 1980; Fox and Dose 1977).

Biblical Apologetics and Engineering

When Martin Gardner described the arguments of young earth advocates in the chapter called "Geology vs. Genesis" of his delightful book *Fads and Fallacies in the Name of Science* (1957), he made no mention of either "scientific" creationism or thermodynamics. Given the hoopla of modern scientific creationism, one might think that its proponents had made some new and momentous discoveries since 1957, thereby rendering Gardner's critique obsolete. Yet many of the arguments advanced by "scientific" creationists are strikingly similar to, if not identical with, those advanced many decades ago by past creationists and exposed by Gardner: for example, creation of an earth with great "apparent age," the alleged circularity of geological dating, the flood theory of fossils, etc. Equally striking is the fact that the early creationists are rarely cited by modern creationists, nor are they generally given credit for the arguments they originated and championed. For this reason many modern proponents of creationism do not realize that the "new" creation science of the 1960s is virtually identical to the biblical apologetics devised in the nineteenth and early twentieth centuries to counter Darwin's theory. Even the thermodynamic arguments, of which Gardner made no mention, were old. To be sure, they were not worthy of mention: first, because they were absurd, and second, because they had gained virtually no visibility in this country, possibly because of their inherent absurdity. It required the audacity of Henry Morris to change all that.

Modern creationists generally credit Henry Morris with discovering the "contradiction" between thermodynamics and evolution. His arguments were first advanced in *The Genesis Flood* (1961), which he coauthored with J. C. Whitcomb. This work has been identified by Clough (1969), among others, as the seminal work that rekindled modern interest in creationism and that laid the technical and scientific groundwork for the "scientific" creation movement. In fact, on page 222 of *The Genesis Flood,* the thermodynamic arguments against evolution are touted as the scientific basis of creationism. But papers by earlier creationist authors that had mentioned thermodynamics were not alluded to or cited anywhere in *The Genesis Flood.*

In the early 1940s R. E. D. Clark (1943) and E. H. Betts (1944) both published papers entitled "Evolution and Entropy" and were awarded prizes for them by the Victorian Institute, an organization founded to combat Darwinism in Britain (White 1900). Clark and Betts appear to be the first creationists to use thermodynamics, and particularly the second law, as an apologetic tool against evolution. Moreover, these articles covered virtually all the thermodynamic arguments that today's "scientific" creationists attribute to Morris. However, both papers remained in obscurity until 1954 when they were discussed by Bernard Ramm in *The Christian View of Science and Scripture* (1954). It was Ramm's conclusion that "thermodynamics may be fatal to evolution." Then, seven years hence, *The Genesis Flood* appeared with numerous references to Ramm's book, although none in Morris's "pioneering" section on thermodynamics.

It is not surprising that Morris capitalized upon the notion that "thermodynamics may be fatal to evolution." Henry Morris was a trained engineer who received his Ph. D. for work in hydrology and hydraulic engineering from the University of Minnesota. He was also one of the founders of the Creation Research Society. Morris could use his status as an engineer to promote the cause of Christian fundamentalism quite effectively. After all, the science of classical thermodynamics certainly had its roots in engineering, and most engineers are rather well drilled in thermodynamic principles and applications. If evolutionists could be lured somehow into debating thermodynamics, rather than natural selection, fossil records, and other like topics, perhaps the creationist engineers could become acclaimed contributors to the revival of fundamentalism. Thermodynamics could never be easily debated by biologists, anthropologists, and geologists. And as experts in a "hard science" firmly rooted in mathematical rigor (while evolution is less so), engineers spouting the second law (even if wrongly) could gain a definite advantage.

A decade after the publication of Whitcomb and Morris's *The Genesis Flood,* two sister ministries of "scientific creationists," both dominated by engineers, were in operation. These were the Creation Research Society, whose *Quarterly* publishes what the creationists see as their best theoretical, research, and review papers, and the Institute for Creation Research, which houses the writing, speaking, and educational ministries of Christian Heritage College, a fundamentalist Bible college in San Diego.

To a large extent, the creationists' polemics against geologists, paleontologists, and biologists were not taken very seriously as science by most educated persons until "entropy"—a much more effective apologetic—was used. Shrouded in mystique, entropy's potential for misinterpretation is well known even to students and practitioners of thermodynamics. Claude Shannon, the inventor of the uncertainty function in communications engineering and the father of information theory, was advised by the internationally renowned mathematician and scientist Jon Von Neumann to call his new uncertainty function *entropy* for two reasons: "In the first place, your uncertainty function has been used in statistical mechanics under that name, so it already has a name. In the second place, and more important, *no one knows what entropy really is, so in a debate you will always have the advantage*" (italics added). Tribus and McIrvine related this humorous little anecdote in the September 1971 issue of *Scientific American* (p. 180), and within a short time the entropy argument against evolution was among the creationists' favorite debate tools. Soon, D. R. Boylan, the most prestigious engineering educator in the leadership of the creationist movement added his testimonial: "The second law has been particularly helpful in developing an apologetic against abiogenesis . . ." (1977, p. 7).

This is not to imply that engineers had discovered a real weakness in evolution or even that their arguments make sense for hydraulics engineering. Indeed, most engineers will easily recognize the flaws in the entropy arguments of modern creationists. Unfortunately, there has been some reluctance by those most capable of exposing the errors of these arguments to speak out. Some of the most vigorous opposition has come from other creationists—those who *are* competent in the field of thermodynamics and who are appalled that their beliefs are being defended by such faulty apologetics (viz., Cramer 1978).

Concluding Remarks

"Scientific" creationism is an evangelical movement of fundamentalist ministries dedicated not to the advancement of science but to the advancement of biblical inerrancy often at the expense of science. Basically the discourse of scientific creationism is an elaborate but confusing system of apologetics and polemics. It is designed to both defend biblical "truths" and to undermine any scientific facts and theories that contradict creationist interpretations of Scripture. In some cases, Morris and his colleagues use *ad hominem* polemics to attack evolution, as when they assert that Satan is the author of evolutionary theories (see, for example, Morris 1963); more often they employ the apologetic device of obfuscation, especially with regard to the second law.

Because the second law of thermodynamics is nonintuitive and because few people have studied it in depth, it is ideally suited to the apologists' favorite techniques of obscurantism. Moreover, the second law does provide a criterion for determining if certain processes are impossible in nature. Hence, by misinterpreting the second law, whether by ignorance or deliberate deception or both, the creationists are able to convince unwitting audiences that evolution is impossible.

In reality, however, the "uphill" processes associated with life not only are compatible with entropy and the second law, but actually depend on them for the energy fluxes off of which they feed. Numerous other kinds of backward processes in smpler, nonliving systems also proceed in this way and do so in complete accord with the second law. The creationists' second law arguments can only be taken as evidence of their willingness to bear false witness against science itself. It is a sad testimonial to the community of professors, engineers, and scientists that so many have ignored their professional responsibilities in failing to expose the creationist thermodynamics apologetic.

REFERENCES CITED

Bentley, W. A. and Humphreys, W. J. 1931. *Snow crystals.* New York: McGraw-Hill.

Betts, E. H. 1944. Evolution and entropy. *Journal of the Transactions of the Victorian Society* 76:1–18.

Boylan, D. R. 1977. Untitled statement. In *Twenty-one scientists who believe in creation.* 2nd ed. San Diego: Creation-Life Pubs.

Broda, E. 1975. *The evolution of the bioenergetic processes,* New York/Oxford: Pergamon Press.

Bronowski, J. 1970. New concepts in the evolution of complexity. *Zygon* 5:-18–35.

Carnot, Sadi N. L. 1824. On the motive power of fire. As reprinted in *The second law of thermodynamics,* (1976), ed. Joseph Kestin, pp. 16–35. Stroudsburg, Pa.: Hutchinson & Ross.

Clark, R. E. D. 1943. Evolution and entropy. *Journal of the Transactions of the Victorian Society* 75:49–71.

Clausius, Rudolf J. E. 1865. *Pogg. ann. bd.* 125:400; also *Mechanische wärmetheorie* 9:44. As cited by J. W. Gibbs in *The scientific papers of J. Willard Gibbs,* vol. 1., p. 55. New York: Dover (1961).

Clough, C. A. 1969. Eight years after: effect of *The Genesis Flood. Creation Research Society Quarterly,* vol. 6, no. 2, pp. 81–88.

Cramer, J. A. 1978. General evolution and the second law of thermodynamics. In *Origins and change: selected readings from the Journal of the American Scientific Affiliation,* ed. David L. Willis, pp. 32–33. Elgin, Ill. The American Scientific Affiliation.

Dawkins, C. R. 1979. Personal letter to Dr. Vince D'Orazio (16 July).

Feynman, Richard P., Leighton, Robert B., and Sands, Matthew. 1963. The second law. Lecture 44 in *The Feynman Lectures on Physics,* vol. 1., section 44-2, p. 44-3. Reading, Mass.: Addison-Wesley.

Fox, Sidney W. 1980. New missing links: how did life begin? *The Sciences* (January) pp. 18–21.

Fox, Sidney W. and Dose, Klaus. 1977. *Molecular evolution and the origin of life.* Rev. ed. New York: Marcel Dekker.

Gale, George. 1979. *Theory of science,* New York: McGraw-Hill.

Gardner, Martin. 1957. *Fads and fallacies in the name of science.* New York: Dover.

Glansdorff, P. and Prigogine, I., 1971. *Thermodynamic theory of structure, stability and fluctuations.* New York: Wiley Interscience.

Graham, Frank D. 1943. *Audel's pumps, hydraulics, air compressors.* New York: Theo. Audel & Co.

Greene, Arthur M. 1911. *Pumping machinery.* New York: Wiley.

Lepkowski, W. 1979. The social thermodynamics of Ilya Prigogine. *Chemical and Engineering News Analysis* 57:30–33.

Levine, Ira N. 1978. *Physical chemistry.* New York: McGraw-Hill.

Lukas, M. 1980. The world according to Ilya Prigogine. *Quest/80,* vol. 4, no. 10, pp. 15–18, 86–88.

Morris, Henry M. 1963. *The twilight of evolution.* Grand Rapids, Mich.: Baker Book House.

—. 1975. *The troubled waters of evolution.* San Diego: Creation-Life Pubs.

Nicolis, G. and Prigogine, I. 1977. *Self organization in non-equilibrium systems.* New York: Wiley.

Patterson, John W. 1979. Ionic and electronic conduction in nonmetallic phases. In *Corrosion chemistry,* eds. George R. Brubaker and P. Beverly Phipps, pp. 96–125. Washington, D. C.: American Chemical Society.

Perelson, A. T. 1975. Network thermodynamics. *Biophysical Journal* 15:667–85.

Prigogine, Ilya. 1977. Time structure and fluctuations. Nobel lecture in chemistry, reprinted in *Science* 201: 777–85.

—. 1980. *From being to becoming.* San Francisco: Freeman, Cooper and Co.

Prigogine, Ilya, Allen, P. M., and Herman, R. 1977. The evolution of complexity and the laws of nature. In *Goals in a global community,* vol. II, eds. Erwin Laszlo and Judah Bierman. New York/Oxford: Pergamon Press.

Ramm, Bernard. 1954. *The Christian view of science and Scripture.* Grand Rapids, Mich.: Erdmans Pubs.

Schrödinger, Erwin. 1945. *What is life?* New York: Macmillan.

Tolansky, S. 1958. Symmetry of snow crystals. *Nature* 181:256–57.

Tribus, Myron and McIrvine, Edward C. 1971. Energy and information. *Scientific American,* vol. 225, no. 3, pp. 179–88.

Whitcomb, John C. and Morris, Henry M. 1961. *The Genesis flood.* Philadelphia: Presbyterian and Reformed Pub. Co.

White, Andrew D. 1900. *A history of the warfare of science and theology in Christendom.* New York: Appleton.

Zebrowski, Ernest, Jr. 1980. *Practical physics.* New York: McGraw-Hill.

7

Molecular Evidence for Evolution

BY THOMAS H. JUKES

Molecular Evolution

Until about thirty years ago, scientists were able to use only the visible characteristics in their studies of the evolution of organisms. In the early 1950s, biochemists started a revolution in biology by means of research with proteins and nucleic acids. These studies had a great impact upon genetics and, consequently, upon our understanding of evolution. An ever-widening vista has been opened as a result of the discovery of the molecular structure of proteins and, more recently, of the sequence of nucleotide bases in DNA.

A new branch of science, molecular evolution, has come into being. Two of the basic principles on which it is founded are that infrequent and cumulative changes take place in DNA through time, and that one type of change, called a "point mutation," accumulates in DNA at a rate that is roughly constant, about 1 to 2 percent per 5 million years. This phenomenon has allowed us to develop a "molecular evolutionary clock." The clock runs more slowly in regions of DNA that are not readily subject to change.

Informational Macromolecules

Molecules composed of nucleic acids or of proteins are called *informational* because they are made of repeating sequences of different units, just like the letters in the words of a sentence. This first became clear when Sanger (1949; Sanger and Tuppy 1951) discovered that insulin was made of a definite sequence of amino acids placed in a certain order. Insulins from the same species of animal always had the same sequence. Within two years, Watson and Crick (1953) found that the same principle was true for the molecule of DNA. The pairing of the four nucleotide bases contained in DNA—adenine, thymine, cytosine, and guanine (A, T, C, and G)—is regular; A always pairs with T and C with G so that they are opposite to each other in the double strands. Each of the single strands can give rise to a pair of double strands, identical with the original pair. As Watson and Crick observed, this "immediately suggests a possible copying mechanism for the genetic material" (1953, p. 737).

The structure of the molecule had further significance. It was clear that each DNA molecule had been copied from a parent DNA molecule. By following this process backward, through hundreds of millions of years, we would arrive eventually at the very beginnings of life as we know it. Any and all changes that have taken place in the intervening years in DNA molecules must therefore be linked with evolution.

Sequences of the four different units in DNA, as we have noted, carry genetic information; DNA codes for the sequence of amino acids in proteins. DNA nucleotide base sequences are translated into corresponding sequences of the twenty different amino acids in proteins by means of the "genetic code." Three consecutive nucleotides in DNA code for one of the amino acids. Therefore, the sequence of amino acids in human insulin corresponds to a sequence of nucleotides in the gene for human insulin, which has now been identified and analyzed. A portion of this sequence is shown in table 1.

Sanger and his collaborators also found that the amino acid sequences of insulin from various species of animals differed slightly. For example, horse insulin has *glycine* in the ninth position of its amino acid A-chain, but in pig insulin the amino acid in this position is *serine.* This could be explained by a change of a single nucleotide in the DNA of the insulin gene during evolution.

In the 1960s, scientists were not able to locate the gene for insulin

Table 1.
PART OF THE NUCLEOTIDE SEQUENCE IN THE HUMAN INSULIN GENE, ALIGNED WITH THE CORRESPONDING AMINO ACID SEQUENCE IN THE A-CHAIN OF INSULIN

-G-G-C-A-T-T-G-T-G-G-A-A-C-A-A-T-G-C-T-G-T-A-C-C-A-G-C-A-T-C-

Gly - Ile - Val - Glu - Gln - Cys - Cys - Thr - Ser - Ile-
1 5 10

in DNA or to determine the sequence of nucleotides in it. However, it was a straightforward although laborious task to analyze proteins for their amino acid sequences.

Hemoglobin, which can be easily crystalized in pure form from the blood of animals, was one of the first proteins used in such studies. As proteins go, it is a conveniently small molecule for analytical research. There are four subunits or chains in hemoglobin, two of which are identical *alpha* chains, each containing 141 amino acids, and two of which are identical *beta* chains, each with 146 amino acids. The alpha and beta chains can be separated and analyzed. Such study revealed that the alpha chains of human beings and gorillas were almost identical, differing only by a single amino acid. The same was true of the beta chains. Each human alpha chain differed from cattle alpha hemoglobin by about 17 amino acids, corresponding to about 12 percent, and each beta chain by about 24 amino acids, corresponding to about 16 percent. As more information accumulated, it became possible to construct "family trees" for the differences between hemoglobin chains of various vertebrates. *The divergences showed the same relationships as those predicted from the classical taxonomic studies of various animals, studies based on ordinary "visible" characteristics.* This was the start of the science of molecular evolution. It has been expanding ever since.

It was also found that, in *any* vertebrate animal, the alpha and beta chains differ from each other by about 65 percent. When this difference is placed on a time scale calibrated from the fossil record, it becomes apparent that the alpha and beta chains are descended from a single ancestral gene that existed about 600 million years ago. Six hundred million years is the time needed to accumulate the differences that exist in modern alpha and beta chains of vertebrates. These conclusions are summarized in table 2 and figure 1. These data are important because they are *only* comprehensible within an evolutionary framework. Given

Table 2.

AVERAGE AMINO ACID DIFFERENCES IN HEMOGLOBIN CHAINS OF SPECIES FROM VARIOUS CLASSES OF THE VERTEBRATES, COMPARED WITH ESTIMATED TIMES OF DIVERGENCE

Animals	*Differences per 100 codons*		*Average for* $\alpha + \beta$	*Million years*†
	α *chain*	β *chain*		
Placental mammals‡, inter alia (except human versus loris)	16.1	16.7	16.4	100
Kangaroo versus placental mammals	21.7	26.9	24.3	160
Monotremes∫ versus theria	30.5	24.5	27.5	190
Chicken versus mammals	29.6	31.7	30.6	215
Viper versus warm-blooded	39.2			290
Amphibians§ versus terrestrial	46.7	48.9	47.8	380
Bony fish¶ versus tetrapods	49.3	49.6	49.5	400
Shark versus bony vertebrates	57.5	63.8	60.6	545
Shark α versus 7β's*			66.7	625
Shark β versus 9 α's**			65.1	

†Million years since divergence, if it is assumed that carp separation was 400 million years ago. ‡Human, loris, mouse, rabbit, dog, bovine. ∫Echidna, platypus. §Newt α-globin, bullfrog β-globin. ¶Carp and goldfish. *Human, mouse, rabbit, kangaroo, opossum, echidna, frog. **Human, mouse, rabbit, kangaroo, opossum, echidna, chicken, viper, carp.

the operation of a molecular clock, the evolutionary model would predict that closely related organisms will differ *less* in the amino acid sequences of their hemoglobin alpha chains than will distantly related organisms. The same will be true when beta chains are compared with each other (table 2). This is because closely related organisms have had less time to accumulate amino acid substitutions since their divergence from their common ancestor than have distantly related organisms. The evolutionary model also predicts that the alpha and beta portions of any and all hemoglobin molecules should be roughly equally divergent, and this is found to be the case. The difference is always about 60 percent. This is because whatever its specific evolutionary history, the hemoglobin of each organism will have inherited amino acid substitutions accumulated at the same rate over the period of time since the initial divergence of

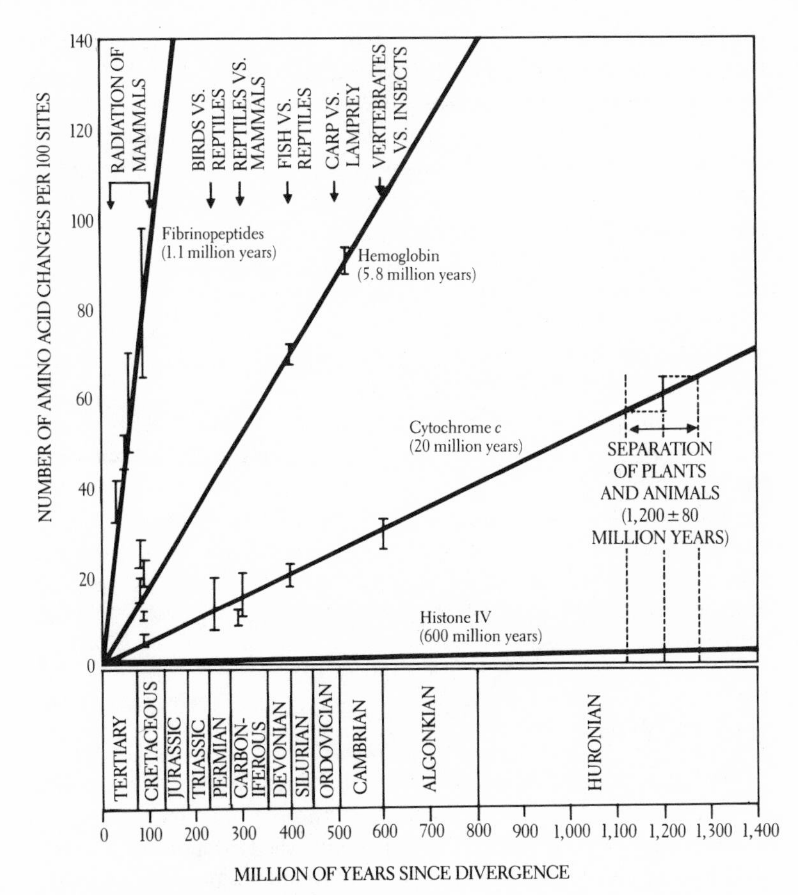

Figure 1. Rates of protein evolution

For each comparison (e.g., fish vs. reptiles, carp vs. lamprey, one mammal group vs. another) the vertical bar gives the experimental scatter of values for the number of amino acid changes per 100 sites for either histone IV, cytochrome *c*, hemoglobin, or fibrinopeptides. The heavy solid lines show the changes in amino acids over time, again for histone IV, cytochrome *c*, hemoglobin, and fibrinopeptides. Time (the horizontal axis) is deduced from the fossil record of points of divergence for the groups of organisms being compared. The rate of change in amino acids for each protein is proportional to the slope of its line. Note that fibrinopeptides change very rapidly and histone IV almost imperceptibly. Indicated in parentheses for each group of proteins is the number of years required to produce a change of 1% in the amino acid sequences of any two diverging lineages. [From Dickerson 1972. Reprinted by permission.]

the *chains,* in this case 600 million years. The upshot is that the hemoglobin molecular data yield independent corroboration of evolution and of the relative evolutionary distances between specific pairs of organisms.

Similar conclusions can be drawn from studying the similarities and differences in the amino acid sequences of *any* protein that is found in a number of different organisms. The only general difference is that some proteins have accumulated amino acid substitutions more slowly than others and are therefore more constant in composition. These differences have a functional basis. For example, the histones are a class of proteins that are bound to DNA in cells that possess a nucleus. They take part in the formation of nucleosomes. Any change in histones could therefore have a destructive effect on the integrity of cells. Thus, the histones of organisms as different as peas and cows are almost identical. In contrast, quickly changing enzymes such as pancreatic ribonuclease, which swims freely in the intestinal juices, also exist; only a small part of the pancreatic ribonuclease molecule takes part in its biological action. Not surprisingly there are big differences between the pancreatic ribonucleases in different species of mammals. We explain this by noting that most of the molecule is free to evolve without restraint.

DNA and Darwin

DNA molecules are subject to constant change. *Mutations* take place in which one nucleotide is replaced by a different one. The amount of DNA in cells can also increase by a process called "duplication." Also, segments of DNA can be switched to different locations on chromosomes. All changes in DNA can potentially produce changes in the characteristics of an organism. However, not all of them do so. We will discuss this later.

All hereditary changes—all changes in DNA—are subject to Darwinian natural selection, a process that weeds out defective organisms. The concept of natural selection is not difficult; Shakespeare alluded to such a concept in *The Merchant of Venice* (act IV, scene 1, line 114):

> *I am a tainted wether of the flock*
> *Meetest for death; the weakest kind of fruit*
> *Drops earliest to the ground; . . .*

Of course, the data of molecular biology allows us to describe the operation of natural selection in much greater detail. Natural selection not only weeds out harmful mutations, it is also an agent of beneficial change. This is a natural consequence of the fact that not all mutations are harmful. A mutation that produces an improvement in some characteristic of a living organism will enable its offspring to compete successfully with other members of the species for survival. Molecular biology provides us with concrete examples of mutational "improvements." Take, for instance, the hemoglobin molecule. Its four units, two alpha and two beta chains, are attracted to each other by binding forces between amino acids at key spots in the hemoglobin chains. This quartet, or "tetramer," has specially advantageous properties for oxygen transport. The tetramer avidly combines with oxygen in the lungs and transports the oxygen to the tissues. When it reaches the tissues the pH is lower, and, in consequence, the tetrameric molecule changes its shape and releases oxygen for use in metabolic processes.

Tetrameric hemoglobin molecules apparently evolved hundreds of millions years ago from hemoglobin molecules that were monomeric and did not form "quartets." Such molecules are found today in a primitive fish, the lamprey, which is regarded as a "living fossil."

We postulate that the first creatures to acquire tetrameric hemoglobin molecules were able to swim faster than their predecessors due to their improved oxygen transport system. They were probably sharks. This interpretation fits nicely with the observation that the alpha and beta chains of hemoglobins have diverged somewhat more than either the alpha chains of sharks and bony vertebrates or the beta chains of sharks and bony vertebrates (table 2). The time of divergence of the alpha and beta chains of hemoglobins seems to have been slightly earlier than the time of divergence of sharks and bony vertebrates.

We explain the interesting case of the lamprey, which does not have a tetrameric hemoglobin, by the fact that it is parasitic on other fish and therefore has not developed a need for swimming rapidly.

Several other families of proteins have been explored with respect to their similarities and differences in different species of living organisms. One such family is that of the cytochromes *c*. These proteins contain between 95 and 105 amino acids. They are found in all eukaryotic organisms, extending from yeasts and molds through insects, green plants, and vertebrate animals. The cytochromes *c* have changed slowly, more slowly than the hemoglobins, because a cytochrome *c* molecule interacts with two other proteins, cytochrome *c-1* and cytochrome oxidase. Such interactions put a strong restraint on evolutionary change in cytochrome *c*.

Evolution of Ribosomes

Ribosomes are present in all living organisms that have an independent existence, covering the whole range from bacteria to vertebrates and green plants. (Viruses do not contain ribosomes; they use the protein-synthesizing mechanisms of their hosts.)

Each ribosome engaged in protein synthesis consists of a large and small subunit. The small subunit contains a single strand of RNA—either 16S or 18S ribosomal RNA, depending on whether it comes from a prokaryote or a eukaryote. Carl Woese (1981) and his colleagues compared the sequence differences of these ribosomal

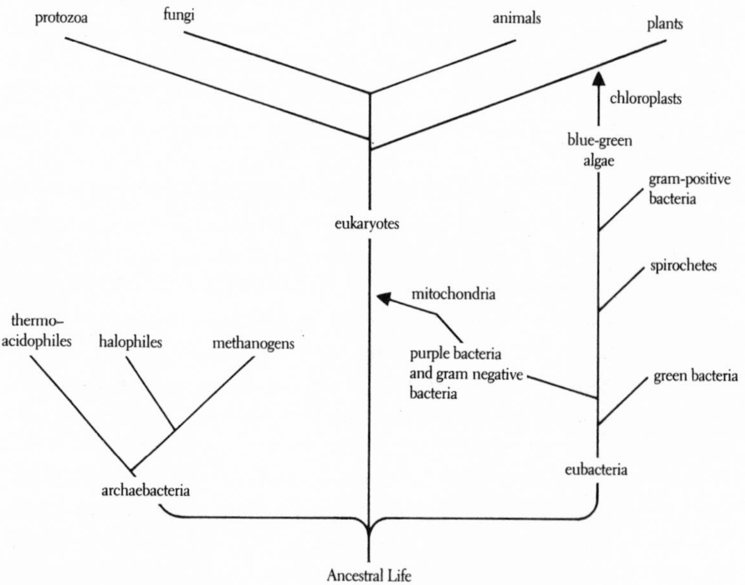

Figure 2. EVOLUTIONARY FAMILY TREE OF TERRESTRIAL LIFE [From Woese 1981.]

RNA molecules from a wide range of different organisms and tabulated the percentage differences in the sequences of these molecules. By this means they showed that there are three rather than two main types of organisms: prokaryotes, eukaryotes, and archaebacteria. Woese's conclusions regarding the evolution of these major groups of organisms are shown in figure 2. Again, he invoked the principle of the "evolutionary clock," based on the observation that changes in DNA, and consequently in RNA and proteins, take place at a roughly uniform rate as measured in nucleotide replacements percent per million years.

Woese's measurements confirmed the postulation that eukaryotic cells were invaded twice by bacteria that took up residence within the cells. The first invasion was by bacteria that became mitochondria. The second event was the entrance of "cyanobacteria." These carried chlorophyll into the primitive cells that later gave rise to all the green plants; the cyanobacteria became the chloroplasts. This is clear because chloroplasts contain ribosomes that can be related by the sequences in their 16S ribosomal RNA to the corresponding ribosomes of modern free cyanobacteria, or blue-green algae. Some of the most primitive fossils that have been found, at 3 billion years old, are cyanobacteria.

Silent Evolutionary Changes in DNA

We have discussed deleterious mutations in DNA; these play an important part in evolution. We have also discussed beneficial or adaptive changes in DNA. Intermediate between these two types of change are *silent* changes, which are neutral or only slightly deleterious. Thus we can perceive a continuous spectrum of changes; mutations are neither all good nor all bad.

An example of a silent change would be a mutation that changed a codon for the amino acid glycine from G-G-A to G-G-G. Both G-G-A and G-G-G code for glycine. This would represent a change in DNA without a corresponding change in the protein. In contrast, if the change were from G-G-A to G-A-A, the codon would now read "glutamic acid"; such a mutation would change the properties of the protein.

In the past few years, large numbers of silent changes have been discovered in the sequences of nucleotides in DNA molecules.

DNA and Evolution

All hereditary characteristics, all living organisms, depend on information contained in DNA molecules and transmitted through RNA formed by transcription of DNA molecules. The term *transcription* refers to the synthesis of RNA molecules complementary to one of the strands of DNA, so that a DNA sequence such as -A-A-T-G-C-C-A-G-G is transcribed into an RNA sequence -U-U-A-C-G-G-U-C-C- (where U is a nucleotide base, uracil, that replaces thymine in RNA).

All changes that have occurred in evolution have necessarily come about as a result of changes in DNA molecules. DNA molecules are never static. Mutations constantly occur in them, and some of the mutations persist.

As organisms increase in complexity, the DNA content per cell also increases. This does not mean that what we commonly regard as the most highly specialized organisms contain the most DNA. For example, certain amphibians contain more DNA per cell than mammals.

The concept of "selfish DNA" refers to the evolutionary tendency of DNA to increase in content per cell, regardless of the needs of the organism. Selfish DNA contains no genetic information but is nevertheless perpetuated in eukaryotic organisms. One form exists as satellite DNA (consisting of large numbers of short repeated sequences connected end to end), and another form consists of dispersed repetitions. Both the spread and the loss of repeated sequences are thought to occur quite often in evolution. Human DNA contains repeated sequences, three hundred nucleotides in length, interspersed with nonrepeated sequences of two thousand nucleotides. About 3 percent of human DNA consists of the short repeated sequences. Such redundant DNA may be shared by members of large groups of organisms. For example, the repetitive human DNA has been shown to be similar, though not identical, to a "family" of short repeated sequences that occurs in a fairly distant relative, the Bonnet monkey.

Two methods are used to study what has happened to DNA during evolution. The first is to compare the DNA of related species. When DNA is heated gently, its two strands separate from each other. Upon cooling, the single strands find their partners in a solution and become double strands again. The same process of annealing takes place to a considerable extent when DNA single strands from two related organisms, such as cats and tigers, are brought into contact with each other. However, a "hybrid" DNA of this type has a lower melting point than the "pure" DNA of either cats or tigers, and the percentage difference, based on depression of melting point, can be calculated. By this procedure a great deal has been learned about relative similarities of DNA molecules from different species. Inferences drawn from such analyses can be compared with inferences drawn from the fossil record (where the time of *divergence* for pairs of organisms can be estimated). Molecular structure and the fossil record provide independent measures of the degree of relatedness. The remarkable outcome is the extent to which these measures agree.

The melting and annealing procedure is now rapidly being replaced by new methods that allow direct comparisons of the nucleotide sequences in DNA molecules. The actual genes in different but related organisms can be directly compared, thus providing the investigator with absolute measures of genetic similarity.

The second method for studying the evolution of DNA is to examine the sequences within the DNA of *single organisms;* such sequences preserve relics of past evolutionary history. A large amount of new information has recently been obtained by this method. In many cases, two different genes for the same function or enzyme are found within a short distance of each other. In other cases, one member of the pair of duplicate genes is no longer functional, or, as one author put it, it remains as a "rusted, decaying hulk."

Genes from several hemoglobin chains—alpha, beta, gamma, delta and epsilon—have been located on human chromosomes, and their complete sequences have been discovered. The gamma, delta, and epsilon hemoglobin chains are "beta-like" and pair with alpha chains to form tetramers, similar to $\alpha_2\beta_2$ tetramers. The $\alpha_2\gamma_2$ tetramer is present in fetal blood. It is replaced by $\alpha_2\beta_2$ hemoglobin soon after birth takes place, when the gamma gene is "turned off" and the beta gene starts functioning. Epsilon chains are used only in early embryonic life. Delta chains are found as $\alpha_2\beta_2$ tetramers in adult human blood. Human beings have two alpha genes, clustered in chromosome no. 10. Another cluster on a different chromosome, no. 11, contains single genes for beta, delta, and epsilon chains, plus two gamma chain genes and two "pseudo beta" genes that are nonfunctional. The cluster contains about sixty thousand nucleotides, most of which are "spaces" between the genes.

Gaps and mutations accumulated in the pseudogenes during evolution after they became nonfunctional. Pseudobeta genes are also present in lemurs, rabbits, and mice, and pseudoalpha genes in mice. There may be as many as ten pseudobeta genes in mice.

Each hemoglobin gene contains two regions called intervening sequences or "introns." The first, or "short" intron has 116 to 130 nucleotides in each of eight genes of various human, rabbit, and mouse hemoglobins. The second, "long" intron has 850 to 904 nucleotides in human hemoglobin genes, 573 in the rabbit beta gene, and 628 to 653 in mouse beta genes. The introns are snipped out and discarded from RNA that is formed by transcription of the genes in DNA. The three remaining pieces are joined end to end before being used for making hemoglobin molecules. All these tailoring procedures are carried out by special enzymes. The same complicated procedures take place in processing all hemoglobin genes.

Introns are placed at identical locations in all functional hemoglobin genes, with one exception. This is the "leghemoglobin" gene, found in root nodules of leguminous plants, which has three introns. One theory is that the gene was carried into plants from animals by a virus.

In some cases, a gene has duplicated and one member of the pair has retained the original function, while the other member has changed enough to take on a new function, at the same time remaining recognizable as a relative. As an example, the two hormones, secretin and glucagon, have different functions, but their genes are obviously related to each other because they have similar sequences.

Cytochrome *c*

Many thousands of different proteins exist in complex organisms such as vertebrates, flowering plants, or insects. Obviously, not all have been well studied. Some were selected for study because they are found in many species, are small enough to be analyzed easily, and are not difficult to purify. One such protein is cytochrome *c*. Cytochrome *c* can be isolated as a pure reddish crystalline substance. Its molecule contains about one hundred amino acids, and a heme group containing iron is attached to two of them. The iron atom functions as an electron donor and acceptor in the oxidative chains of processes taking place in living cells. Cytochrome *c* molecules are less variable than hemoglobin molecules; for example, sheep, pigs, and cattle have identical molecules of cytochrome *c*. Also, the differences between cytochrome *c* molecules are small even when one compares those obtained from species as diverse as horses, yeast cells, wheat plants, and fruit flies. The odds against this similarity being due to chance are overwhelmingly great.

Because different molecules accumulate amino acid changes at different rates, some allow us to explore relationships among distantly related organisms, while others yield the resolution necessary to explore relationships among closely related organisms. Comparison of the sequences of different hemoglobins was useful in constructing a "family tree" for all vertebrates. Comparisons of slowly evolving cytochromes *c* enable us to construct such a tree for all eukaryotic organisms, including vertebrates, green plants, yeasts, insects, and protozoa. As expected from an evolutionary model, the family tree of cytochrome sequences resembles quite closely the evolutionary scheme postulated by biologists long before any cytochrome *c* sequences were known.

The rate of change of any molecule is not invariant. There are minor, but interesting, differences in the rate of evolution of cytochrome *c* as calculated using the fossil record for different animals. For example, the evolutionary changes of cytochrome *c* in rattlesnakes were considerably more rapid than in snapping turtles (Jukes and Holmquist 1972). Such deviations smooth out when large groups of organisms are considered and in no way jeopardize the general conclusions about evolutionary relationships drawn from the study of molecular similarities and differences.

Dickerson (1972) has provided a thorough and lucid discussion of the evolutionary relationship among cytochromes *c* from various species of living organisms. Between 1.5 and 2 billion years ago, when the living organisms on the earth were single-celled, a biochemical change took place that enabled energy to be produced by combining the breakdown products of food with molecules of oxygen. Cytochrome *c* was one of the substances that made this possible, and its descendants can be found today in every living cell that has a nucleus. This protein usually has 104 amino acids, although forms of the molecule found in some organisms may be slightly longer, or, more rarely, slightly shorter. Cytochrome *c* has evolved so slowly that in human beings and in the mold *Neurospora crassa* 60 of the 104 amino acid sites are identical, and the three-dimensional structures of the two molecules are practially identical.

Attached to the central region of cytochrome *c* is an atom of iron, surrounded by pyrrole rings. The iron atom functions as a carrier of electrons to transfer energy in metabolic chemistry.

Crystallographic studies have shown that cytochromes *c* always have the same three-dimensional structure. The maintenance of this structure throughout evolutionary history has had a strong bearing upon the stability of the amino acid composition of the cytochromes *c;* many of the amino acids have not changed during more than a billion years of evolution. Other amino acids have undergone two or three changes in which they have been replaced by a chemically similar amino acid, and a few amino acids, on the outside of the molecule, where changes seem to be very easily tolerated, have undergone from five to nine evolutionary changes.

All this fits with the evolutionary prediction that changes constantly occur in DNA during evolution, but only those changes that can be accommodated will persist. Therefore changes, or mutations, in the cytochrome *c* gene have a lethal effect if they radically change the three-dimensional structure of cytochrome *c* so that it cannot carry out its function of transferring electrons.

The family of cytochrome *c* molecules provides one of the best examples of the "molecular evolutionary clock," in which the differences between a protein as it occurs in two different species of organisms accumulate at an approximately constant rate, as shown by the fossil record and by isotopic dating of rocks. The rate of evolution of cytochrome *c,* compared with several other proteins, is shown in figure 1. About 20 million years are required to produce a change of 1 percent in the amino acid sequence of any two diverging lines of cytochrome *c.* This enables an estimate of a date of 1.2 billion years ago for the divergence of flowering plants and animals! The similarity between the cytochrome *c* of human beings and that of wheat plants leads to the conclusion that these two species have descended from a common ancestor that lived about 1.2 billion years ago.

Given the slow rate of change of cytochrome *c,* it should come as no surprise that the cytochrome *c* molecules in humans are identical with those of chimpanzees. Indeed, the faster evolving hemoglobins of humans and chimpanzees are also identical. A close relationship between humans and apes was postulated many years ago, based on the evidence of the classical fields of comparative anatomy and embryology. The data of molecular biology has served only to confirm this conclusion. It is a conclusion that is rejected by creationists, no matter how conclusive the evidence. Today's protests differ little from those of the Scopes era, when, for example, Dr. Kofoid, professor of zoology at the University of California, Berkeley, touched on the striking similarities in the skeletal anatomies of humans and apes:

> Dr. Kofoid had been asked to address the Ministerial Association of Los Angeles on the subject of evolution. Illustrating his speech by drawings and photographs, he had just pointed out the marked skeletal similarity between the anthropoid apes and man, when the entire clerical audience burst into the noisy singing of a hymn. This continued until Dr. Kofoid was completely drowned out and obliged to leave the platform to which he had been invited.
>
> [Shipley 1927, p. 270]

That was 1926. Now let's examine the protests of modern day creationists as they attempt to discredit the conclusions that biologists have drawn from the data of molecular biology.

Molecular Evolution and "Scientific Creationism"

In order to maintain that the earth is ten thousand years old and to deny common ancestry for organisms in general (and for apes and humans in particular), creationists must totally reject the overwhelming and fascinating evidence for evolution that has emerged from the study of molecular biology. They resort, instead, to an odd critique of these data. The creationists have no problem in understanding that mutations can be deleterious and may result in the death of living organisms. But they cannot understand that, occasionally, a mutation may be beneficial. This fact is the essence of Darwinian evolution, and it is a logical counterpart of the observation that most mutations are harmful. Creationists are also prone to making unsupportable statements about molecular biology. For example, they allege that ribosomes have not evolved but are the product of special creation. On the contrary, evolution taking place in ribosomes is the key to understanding evolution from some of its earliest beginnings.

Two creationists, Kofahl and Segraves published a book called *The Creation Explanation* (1975) that specifically challenges the article by Dickerson (1972), summarized above. (Presumably, the section in which the challenge was made was written by Kofahl, since Segraves has no scientific training.)

On pages 167 and 168 of their book, Kofahl and Segraves document the structure of cytochrome *c* by reproducing tables I and II from Dickerson's article. They state that "the segment from position 70 to 80 is invariant, for it is the heart of the active center of the enzyme molecule," and that "there is no evidence for the evolutionary development of the overall structure *nor of the active center* . . . Thus the evidence supports the view that cytochrome *c* was designed, not evolved" (1975, p. 167, italics added). The constancy of residues 70 and 80 from most species is well known, and the evolutionary interpretation is, of course, that mutations in this region are lethal and hence do not persist. Kofahl and Segraves explain the constancy of this region by saying that "cytochrome *c* was designed, not evolved" (p. 167), implying that the constancy of this polypeptide sequence of eleven amino acids was a decision specifically made by God and that "the invariance of all the cytochrome c's in their critical parts is evidence for intellgent, purposeful design" (p. 166). Why constancy of structure should be evidence of design rather than stability mediated by natural selection is not clear, but if Kofahl and Segraves mean to imply that variability of this segment would constitute falsification of their hypothesis of design, then their hypothesis must be taken as having been falsified.

Indeed, all we need do is go back further in evolutionary time, to protozoa, and the constancy of residues 70 to 80 disappears:

	70										80
Vertebrates and green plants	-Asn	-Pro	-Lys	-Lys	-Tyr	-Ile	-Pro	-Gly	-Thr	-Lys	-Met-
Crithidia	-Asn	-Pro	-Lys	-Lys	-Phe	-Met	-Pro	-Gly	-Thr	-Lys	-Met-
Euglena	-Asn	-Pro	-Lys	-Lys	-Tyr	-Val	-Pro	-Gly	-Thr	-Lys	-Met-

Going back still further, the cytochrome c_2 of *Rhodospirillum rubrum,* a photosynthetic bacterium, has an addition in the middle (shown italicized below) as follows:

R. rubrum

73
-Asn -Pro -Lys -Ala -Phe -Val -Leu -Glu -*Lys* -*Ser* -*Gly* -*Asp* -*Pro* -*Lys* -*Ala* -*Lys*
91
-Ser -Lys -Met-

Vertebrate

70
-Asn -Pro -Lys -Lys -Tyr -Ile -Pro -Gly -
80
-Thr -Lys -Met

Thus the conclusion by Kofahl and Segraves is based on an assertion of an invariance that does not exist. All that is evident is that evolution of this sequence takes place very slowly. Clearly, residues 70 to 80 do not change *unless* and *until* many other evolutionary changes have occurred. Why not conclude that the Creator planned that evolution could take place, rather than concerning Himself with the invariance of amino acids 70 to 80 in the cytochromes *c* of some, but not all, eukaryotes?

Kofahl and Segraves then state:

> As reported in Dr. Dickerson's article, this time [required for a change of 1% in the amino acid chain] is about 20 million years between groups. . . . But . . . such values within the vertebrate group vary all the way from seven million years to 50 million years for this selected set of pairs. . . . The spread of values is so wide that it can hardly be said that the data offers [*sic*] good support for the theory that the vertebrates are related by evolution and that the rate of change of the cytochrome *c* molecule was constant with time within the vertebrate group. [1975, p. 166]

This erroneous conclusion shows a lack of comprehension of Dickerson's article. The vertebrate group includes bony fishes, cartilaginous fishes, amphibians, reptiles, birds, and mammals. As stated above, it is well known that the rate of evolution of cytochrome *c* does vary in different species of vertebrates, as, for example, turtles and rattlesnakes (Jukes and Holmquist 1972). But when cytochromes *c* are considered as a family of molecules common not only to vertebrates but to more distantly related species such as insects, molds, and green plants, these minor perturbations in evolutionary rate "smooth out."

Of course, different families of molecules can show markedly different rates of change, as pointed out above. Histones from various eukaryotic organisms show less variation than any other class of proteins that has been examined. This is explained by their function; histones bind to DNA in the chromosomes. We have here a control process that is essential to the genetic mechanism. Not surprisingly, the molecules of histones 4 from peas and cattle differ in only 2 of their 102 amino acids. However, when the *genes* for the histones of different organisms (such as two genera of sea urchins) are compared, they contain a large number of differences. But almost all of the differences are "silent" changes in the third position of codons, and the amino acid sequences are virtually unchanged (table 3).

It seems clear that nucleotide substitutions in the genes for histones have taken place continually and at an approximately constant rate. It seems also clear, using the case of the histones as an example, that substitutions that do not affect the amino acid composition of a protein are tolerated, but amino acid–altering changes are accepted only if they do not impair the function of the protein.

Contrasting sharply with the slowly evolving histones are the fibrinopeptides. These are short proteinlike molecules containing about twenty amino acids. They are cut out of fibrinogen and discarded during the process of blood clotting, when fibrinogen is converted to fibrin. Because they are discarded there appears to be very little constraint upon their composition; thus they evolve very rapidly (fig. 1).

RNA nucleotide sequences can also change rapidly. Such change nullifies one of the creationists' favorite arguments. Creationists frequently state that evolution will always remain an "untestable" theory

Table 3.
RELATION BETWEEN NUCLEOTIDE SUBSTITUTIONS AND AMINO ACID REPLACEMENTS IN COMPARISON OF HISTONES 2A, 2B, 3, AND 4 GENES IN TWO SEA URCHINS, *STRONGYLOCENTRUS PURPURATUS* AND *PSAMMECHINAS MILIARIS*

Nucleotides compared	1,212
Nucleotide substitutions	161
Silent nucleotide substitutions	148
Amino acid replacements	11
Silent substitutions as percent of all nucleotide substitutions	92

because evolution cannot be shown to take place within the context of actual laboratory research. Evolution is said to proceed on a time scale based on units of 1 million years. Human experience in evolutionary biology covers a period of less than 200 years.

Recently the H3 hemagglutinin genes of different strains of influenza viruses were sequenced. This is an RNA virus that evolves very rapidly. Two strains of Hong Kong influenza virus were compared, and each contained 1,701 nucleotide sites. Sixty-three sites were occupied by different nucleotides, and 34 of those represented silent substitutions. There were twenty-eight amino acid replacements. This pattern of change is comparable to that encountered in hemoglobin genes of species that shared a common ancestor about 30 to 50 million years ago. However, the time of divergence of the influenza strains was estimated as only 7 years (Verhoeyen et al. 1980). Here is truly an example of "evolution in action."

Metabolic Pathways

Kofahl and Segraves state that the existence of alternate metabolic pathways is "often difficult to explain from the evolutionary point of view" (1975). Actually, no such difficulties exist. For example, the cells of green plants contain alternate metabolic pathways for the biosynthesis of amino levulinic acid (ALA), a precursor of the pyrrole ring present in chlorophyll and heme. The first pathway, which is common to plants, animals, and many bacteria, is via a reaction involving glycine and succinyl-CoA. The second, occurring in chloroplasts, uses either α-ketoglutarate or glutamate as starting material (Weeden 1981), and the second pathway is the major route to ALA in cyanobacteria. These lack succinyl-CoA synthetase. The evolutionary explanation is that chloroplasts are descendants of cyanobacteria (blue-green algae) that infected the evolutionary predecessors of the cells of green plants and became chloroplasts. There are other pairs of alternate pathways in green plants that show the same pattern of relationships, namely, one of the two pathways is present in chloroplasts and is also found in cyanobacteria. Examples of these are the biosynthesis of proline and other amino acids. According to Kofahl and Segraves, "Such variations illustrate the infinite wisdom and power of the Creator in His capacity to design and create a seemingly unlimited number of working life systems. . . ." (1975,

p. 169). But such variations fit well with the expected pattern of evolution and are inherent in its processes. Scientific discussion is obviously impossible with Kofahl and Segraves because they invariably fall back on the same "explanation" for all natural phenomena, namely, that they are products of special creation.

Furthermore, the creationists do not hesitate to take on themselves the authority to decide cosmological questions. For example, Morris (1972) informs us that there are positively no manlike intelligences living elsewhere in the universe. We may contrast this attitude with a statement by Father Theodore M. Hesburgh, president, University of Notre Dame:

> Just last week, I was discussing the subject [of extraterrestrial intelligence] with a Russian lawyer who regarded me with some surprise and asked: "Surely you must abandon your theology when you consider these possibilities?" "Indeed, I don't," I replied. "It is precisely because I believe theologically that there is a being called God, and that He is infinite in intelligence, freedom and power, that I cannot take it upon myself *to limit what He might have done.*" Once He created the Big Bang—and there had to be something, call it energy, hydrogen, or whatever, to go bang—He could have envisioned it going in billions of directions as it evolved, including billions of life forms and billions of kinds of intelligent beings. I will go even further. There conceivably can be billions of universes created with other Big Bangs or different arrangements. Why limit infinite Power or Energy which is a name of God? We should get some hint from the almost, but not quite, infinite profusion of the Universe we still know only in part. Only one consideration is important here regarding creation. Since God is intelligent, however He creates—"Let there be light"—Bang—or otherwise, whatever He creates is a cosmos and not a chaos since all His creation has to reflect Him. What reflects Him most is intelligence and freedom, not matter. "We are made in His image," why suppose that He did not create the most of what reflects Him the best? He certainly made a lot of matter. Why not more intelligence, more free beings, who alone can seek and know Him?
>
> [Morrison, Billingham, Wolfe 1977, p. vii]

Kofahl and Segraves's "creation model", into which "the data of the sciences can be fitted," contains the following provisions: plant life was created before the sun, moon, and stars were "either created or brought into a condition which made them suitable for telling time, seasons, years and days"; "one man was first created . . . and from his side Eve was created, so that the entire race generated from them would be one in Adam"; in Eden before the fall, "man and animals lived by eating plants as food. . . . The tree of life was supplied to enable man to live forever"; when sin entered the scene, not only all men but "the entire plant and animal creation was dragged into a parallel downward course, a path not of evolution, but of devolution"; "finally, God had to judge and cleanse the world. One godly family was chosen by God to preserve the race. By the time the Ark was completed and the Flood came, only eight believed strongly enough to break with the world system and go God's way. The remainder of the race perished, along with the air-breathing animals. . . ."; after the flood, "the ark, a 30,000-ton vessel some 450 feet long, disgorged its precious cargo of animals and humans which overspread the ruined earth's surface, taking directions determined by the providence of a sovereign Creator. . . . This explains how . . . most of the marsupials such as kangaroos [are found] in or near Australia"; this creation model may "serve as a guide to Christians who desire to pursue scientific research in channels which will be blessed by God because His truth has been accepted as the foundation for expanding human knowledge of His creation" (1975, pp. 221–29).

I believe that these statements by Kofahl and Segraves are sufficient to illustrate the fact that scientific discourse or argument with the creationists is an exercise in complete futility. Their attempt to inflict their rigid dogma upon the educational system of the United States, if successful, would mean the extinction of science and a return to the Dark Ages.

REFERENCES CITED

Dickerson, Richard E. 1972. The structure and history of an ancient protein. *Scientific American,* vol. 226, no. 4, pp. 58–72.

Jukes, Thomas H. and Holmquist, W. Richard. 1972. Evolutionary clock: non-constancy of rate in different species. *Science* 177:530–32.

Kofahl, Robert E. and Segraves, Kelley L. 1975. *The creation explanation.* Wheaton, Ill.: Harold Shaw Publ.

Morris, Henry M. 1972. *The remarkable birth of planet earth.* Minneapolis, Minn.: Dimension Books.

Morrison, Philip, Billingham, John, and Wolfe, John, eds. 1977. *The search for extraterrestrial intelligence.* Washington, D. C.: NASA, U. S. Government Printing Office.

Sanger, F. 1949. Fractionation of oxidized insulin. *Biochemical Journal* 44:-126–28.

Sanger, F. and Tuppy, H. 1951. The amino acid sequence in the phenylalanyl chain of insulin. *Biochemical Journal* 49:481–86.

Shipley, Maynard. 1927. *The war on modern science.* New York: Knopf.

Verhoeyen, Martine, Fang, Rongxiang, Jou, Willy M., Devos, René, Huylebroeck, Danny, Saman, Eric, and Fiers, Walter. 1980. Antigenic drift between the haemagglutinin of the Hong Kong influenza strains A/Aichi/2/68 and A/Victoria/3/75. *Nature* 286:771–76.

Watson, James D. and Crick, Francis H. C. 1953. Molecular structure of nucleic acids. *Nature* 171:737–38.

Weeden, N. F. 1981. Genetic and biochemical implications of the endosymbiotic origin of the chloroplast. *Journal of Molecular Evolution* 17:133–39.

Woese, Carl R. 1981. Archaebacteria. *Scientific American,* vol. 244, no. 6, pp. 98–122.

8

Darwin's Untimely Burial —Again!*

BY STEPHEN JAY GOULD

Introduction

I wrote "Darwin's Untimely Burial" seven years ago in response to an article in *Harper's Magazine* by journalist Tom Bethell entitled "Darwin's Mistake," (1976). Bethell had raised again one of the oldest (and most rotten) chestnuts in the panoply of anti-Darwinian (not antievolutionary) arguments—the claim that natural selection is a meaningless tautology. The simple argument, refuted again and again since it first arose during the nineteenth century, holds that Darwinians define fitness in terms of survival. Therefore, the phrase, "survival of the fittest" reduces to "survival of those that survive"—clearly a meaningless claim. But, as I demonstrate in the article, this argument is based on a false premise and fundamental misunderstanding of Darwin's argument. Darwin used a criterion of fitness—good design in the engineering sense

*This essay is reprinted with the permission of the author and of *Natural History,* in which it first appeared in October 1976, on pp. 24–30. Page numbers of quotations have been added for this volume.

—that is both prior to and independent of mere survival.

Since America's leading creationists are singularly devoid of shame in their willingness to cite any antievolutionary argument that anybody has ever made, no matter how tenuous or demonstrably false, I was not surprised to find the old "tautology" claim circulating prominently in their critiques. Both leading "intellectuals" and public debaters of the Institute for Creation Research have raised it recently. In No. 51 of their Impact Series (1977), Henry Morris writes: "There are now many evolutionists who recognize that the 'theory of evolution' is really a tautology, with no predictive value." (Please note, however, that the false claim for tautology was advanced only against Darwin's mechanism of natural selection, not against the idea of evolution itself.) In an address entitled "Creation, Evolution, and Public Education," Duane Gish cites Bethell's article and writes:

> It is an astounding fact that while at the time Darwin popularized it, the concept of natural selection seemed to explain so much, today there is a growing realization that the presently accepted concept of natural selection really explains nothing. It is a mere tautology, that is, it involves circular reasoning. . . . When it is asked, what survives, the answer is, the fittest. But when it is asked, what are the fittest, the answer is, those that survive! Natural selection thus collapses into a tautology, devoid of explanatory value. [1976, p. 6]

In this light, Laurie Godfrey, the editor of this volume, asked for permission to reprint my rebuttal to Bethell and defense of natural selection. Old arguments may be like old soldiers, and one can only hope that a good push might speed the process of fading away.

Darwin's Untimely Burial

In one of the numerous movie versions of *A Christmas Carol,* Ebenezer Scrooge encounters a dignified gentleman sitting on a landing as he mounts the steps to visit his dying partner, Jacob Marley. "Are you the doctor?" Scrooge inquires. "No," replies the man, "I'm the undertaker; ours is a very competitive business." The cutthroat world of intellectuals must rank a close second, and few events attract more notice than a proclamation that popular ideas have died. Darwin's theory of natural selection has been a perennial candidate for burial. Tom Bethell held the most recent wake in a piece called "Darwin's Mistake" (1976): "Darwin's theory, I believe, is on the verge of collapse" (p. 72). "Natural selection was quietly abandoned, even by his most ardent supporters, some years ago" (p. 74). News to me, and I, although I wear the Darwinian label with some pride, am not among the most ardent defenders of natural selection. I recall Mark Twain's famous response to a premature obituary: "The reports of my death are greatly exaggerated."

Bethell's argument has a curious ring for most practicing scientists. We are always ready to watch a theory fall under the impact of new data, but we do not expect a great and influential theory to collapse from a logical error in its formulation. Virtually every empirical scientist has a touch of the Philistine. Scientists tend to ignore academic philosophy as an empty pursuit. Surely, any intelligent person can think straight by intuition. Yet Bethell cites no data at all in sealing the coffin of natural selection, only an error in Darwin's reasoning: "Darwin made a mistake sufficiently serious to undermine his theory. And that mistake has only recently been recognized as such. . . . At one point in his argument, Darwin was misled" (p. 72).

Although I will try to refute Bethell, I also deplore the unwillingness of scientists to explore seriously the logical structure of arguments. Much of what passes for evolutionary theory is as vacuous as Bethell claims. Many great theories are held together by chains of dubious metaphor and analogy. Bethell has correctly identified the hogwash surrounding evolutionary theory. But we differ in one fundamental way: for Bethell, Darwinian theory is rotten to the core; I find a pearl of great price at the center.

Natural selection is the central concept of Darwinian theory—the fittest survive and spread their favored traits through populations. Natural selection is defined by Spencer's phrase "survival of the fittest," but what does this famous bit of jargon really mean? Who are the fittest? And how is "fitness" defined? We often read that fitness involves no more than "differential reproductive success"—the production of more surviving offspring than other competing members of the population. Whoa! cries Bethell, as many others have before him. This formulation defines fitness in terms of survival only. The crucial phrase of natural selection means no more than "the survival of those who survive"—a vacuous tautology. (A tautology is a phrase—like "my father is a man" —containing no information in the predicate ["a man"] not inherent in the subject ["my father"]. Tautologies are fine as definitions, but not as testable scientific statements—there can be nothing to test in a statement true by definition.)

But how could Darwin have made such a monumental, two-bit mistake? Even his severest critics have never accused him of crass stupidity. Obviously, Darwin must have tried to define fitness differently—to find a criterion for fitness independent of mere survival. Darwin did propose an independent criterion, but Bethell argues quite correctly that he relied upon analogy to establish it, a dangerous and slippery strategy. One might think that the first chapter of such a revolutionary book as *Origin of Species* would deal with cosmic questions and general concerns. It doesn't. It's about pigeons. Darwin devotes most of his first forty pages to "artificial selection" of favored traits by animal breeders. For here an independent criterion surely operates. The pigeon fancier knows what he wants. The fittest are not defined by their survival. They are, rather, allowed to survive because they possess desired traits.

The principle of natural selection depends upon the validity of an analogy with artificial selection. We must be able, like the pigeon fancier, to identify the fittest beforehand, not only by their subsequent survival. But nature is not an animal breeder; no preordained purpose regulates the history of life. In nature, any traits possessed by survivors must be counted as "more evolved"; in artificial selection, "superior" traits are defined before breeding even begins. Later evolutionists, Bethell argues, recognized the failure of Darwin's analogy and redefined "fitness" as mere survival. But they did not realize that they had undermined the logical structure of Darwin's central postulate. Nature provides no independent criterion of fitness; thus, natural selection is tautological.

Bethell then moves to two important corollaries of his major argument. First, if fitness only means survival, then how can natural selection be a "creative" force, as Darwinians insist. Natural selection can only tell us how "a given type of animal became more numerous"; it cannot explain "how one type of animal gradually changed into another." Secondly, why were Darwin and other eminent Victorians so sure that mindless nature could be compared with conscious selection by breeders.

Bethell argues that the cultural climate of triumphant industrial capitalism had defined any change as inherently progressive. Mere survival in nature could only be for the good: "It is beginning to look as though what Darwin really discovered was nothing more than the Victorian propensity to believe in progress" (p. 72).

I believe that Darwin was right and that Bethell and his colleagues are mistaken: criteria of fitness independent of survival can be applied to nature and have been used consistently by evolutionists. But let me first admit that Bethell's criticism applies to much of the technical literature in evolutionary theory, especially to the abstract mathematical treatments that consider evolution only as an alteration in numbers, not as a change in quality. These studies do assess fitness only in terms of differential survival. What else can be done with abstract models that trace the relative successes of hypothetical genes A and B in populations that exist only on computer tape? Nature, however, is not limited by the calculations of theoretical geneticists. In nature, A's "superiority" over B will be *expressed* as differential survival, but it is not *defined* by it—or, at least, it better not be so defined, lest Bethell et al. triumph and Darwin surrender.

My defense of Darwin is neither startling, novel, nor profound. I merely assert that Darwin was justified in analogizing natural selection with animal breeding. In artificial selection, a breeder's desire represents a "change of environment" for a population. In this new environment, certain traits are superior a priori: (they survive and spread by our breeder's choice, but this is a *result* of their fitness, not a definition of it). In nature Darwinian evolution is also a response to changing environments. Now, the key point: certain morphological, physiological, and behavioral traits should be superior a priori as designs for living in new environments. These traits confer fitness by an engineer's criterion of good design, not by the empirical fact of their survival and spread. It got colder before the woolly mammoth evolved its shaggy coat.

Why does this issue agitate evolutionists so much? OK, Darwin was right: superior design in changed environments is an independent criterion of fitness. So what. Did anyone ever seriously propose that the poorly designed shall triumph? Yes, in fact, many did. In Darwin's day, many rival evolutionary theories asserted that the fittest (best designed) must perish. One popular notion—the theory of racial life cycles—was championed by a former inhabitant of the office I now occupy, the great

American paleontologist Alpheus Hyatt. Hyatt claimed that evolutionary lineages, like individuals, had cycles of youth, maturity, old age, and death (extinction). Decline and extinction are programmed into the history of species. As maturity yields to old age, the best-designed individuals die and the hobbled, deformed creatures of phyletic senility take over. Another anti-Darwinian notion, the theory of orthogenesis, held that certain trends, once initiated, could not be halted, even though they must lead to extinction caused by increasingly inferior design. Many nineteenth-century evolutionists (perhaps a majority) held that Irish elks became extinct because they could not halt their evolutionary increase in antler size; thus, they died—caught in trees or bowed (literally) in the mire. Likewise, the demise of saber-toothed "tigers" was often attributed to canine teeth grown so long that the poor cats couldn't open their jaws wide enough to use them.

Thus, it is not true, as Bethell claims, that any traits possessed by survivors must be designated as fitter. "Survival of the fittest" is not a tautology. It is also not the only imaginable or reasonable reading of the evolutionary record. It is testable. It had rivals that failed under the weight of contrary evidence and changing attitudes about the nature of life. It has rivals that may succeed, at least in limiting its scope.

If I am right, how can Bethell claim, "Darwin, I suggest, is in the process of being discarded, but perhaps in deference to the venerable old gentleman, resting comfortably in Westminster Abbey next to Sir Isaac Newton, it is being done as discreetly and gently as possible with a minimum of publicity" (p. 75). I'm afraid I must say that Bethell has not been quite fair in his report of prevailing opinion. He cites the gadflies C. H. Waddington and H. J. Muller as though they epitomized a consensus. He never mentions the leading selectionists of our present generation—E. O. Wilson or D. Janzen, for example. And he quotes the architects of neo-Darwinism—Dobzhansky, Simpson, Mayr, and J. Huxley—only to ridicule their metaphors on the "creativity" of natural selection. (I am not claiming that Darwinism should be cherished because it is still popular; I am enough of a gadfly to believe that uncriticized consensus is a sure sign of impending trouble. I merely report that, for better or for worse, Darwinism is alive and thriving, despite Bethell's obituary.)

But why was natural selection compared to a composer by Dobzhansky; to a poet by Simpson; to a sculptor by Mayr; and to, of all people,

Mr. Shakespeare by Julian Huxley? I won't defend the choice of metaphors, but I will uphold the intent, namely, to illustrate the essence of Darwinism—the creativity of natural selection. Natural selection has a place in all anti-Darwinian theories that I know. It is cast in a negative role as an executioner, a headsman for the unfit (while the fit arise by such non-Darwinian mechanisms as the inheritance of acquired characters or direct induction of favorable variation by the environment). The essence of Darwinism lies in its claim that natural selection creates the fit. Variation is ubiquitous and random in direction. It supplies the raw material only. Natural selection directs the course of evolutionary change. It preserves favorable variants and builds fitness gradually. In fact, since artists fashion their creations from the raw material of notes, words, and stone, the metaphors do not strike me as inappropriate. Since Bethell does not accept a criterion of fitness independent of mere survival, he can hardly grant a creative role to natural selection.

According to Bethell, Darwin's concept of natural selection as a creative force can be no more than an illusion encouraged by the social and political climate of his times. In the throes of Victorian optimism in imperial Britain, change seemed to be inherently progressive; why not equate survival in nature with increasing fitness in the nontautological sense of improved design.

I am a strong advocate of the general argument that "truth" as preached by scientists often turns out to be no more than prejudice inspired by prevailing social and political beliefs. I have devoted several essays to this theme because I believe that it helps to "demystify" the practice of science by showing its similarity to all creative human activity. But the truth of a general argument does not validate any specific application, and I maintain that Bethell's application is badly misinformed.

Darwin did two very separate things: he convinced the scientific world that evolution had occurred and he proposed the theory of natural selection as its mechanism. I am quite willing to admit that the common equation of evolution with progress made Darwin's first claim more palatable to his contemporaries. But Darwin failed in his second quest during his own lifetime. The theory of natural selection did not triumph until the 1940s. Its Victorian unpopularity, in my view, lay primarily in its denial of general progress as inherent in the workings of evolution. Natural selection is a theory of *local* adaptation to changing environments. It proposes no perfecting principles, no guarantee of general improvement; in short, no reason for general approbation in a political climate favoring innate progress in nature.

Darwin's independent criterion of fitness is, indeed, "improved design," but not "improved" in the cosmic sense that contemporary Britain favored. To Darwin, improved meant only "better designed for an immediate, local environment." Local environments change constantly: they get colder or hotter, wetter or drier, more grassy or more forested. Evolution by natural selection is no more than a tracking of these changing environments by differential preservation of organisms better designed to live in them: hair on a mammoth is not progressive in any cosmic sense. Natural selection can produce a trend that tempts us to think of more general progress—increase in brain size does characterize the evolution of group after group of mammals. But big brains have their uses in local environments; they do not mark intrinsic trends to higher states. And Darwin delighted in showing that local adaptation often produced "degeneration" in design—anatomical simplification in parasites, for example.

If natural selection is not a doctrine of progress, then its popularity cannot reflect the politics that Bethell invokes. If the theory of natural selection contains an independent criterion of fitness, then it is not tautological. I maintain, perhaps naïvely, that its current, unabated popularity must have something to do with its success in explaining the admittedly imperfect information we now possess about evolution. I rather suspect that we'll have Charles Darwin to kick around for some time.

REFERENCES CITED

Bethell, Thomas. 1976. Darwin's mistake. *Harper's Magazine,* vol. 252, no. 1509, pp. 70–75.

Gish, Duane. 1976. Creation, evolution and public education. San Diego: Institute for Creation Research.

Morris, Henry. 1977. The religion of evolutionary humanism and the public schools. ICR *Impact* Series, no. 51, pp. i–iv.

9

The Geological and Paleontological Arguments of Creationism

BY DAVID M. RAUP

Introduction

This essay is inspired by my rereading of "Science and Creation" by Boardman, Koontz, and Morris (1973). "Science and Creation" is not the most recent treatment, but it is one of the best in the sense of being a clear and unambiguous statement of the case made by contemporary creationists against the conventional wisdom of evolution. The same basic ground is covered by later books such as those by Wysong (1976) and Gish (1978). I will be concerned here only with the strictly geological and paleontological parts of the argument. Gish (1978) has popularized the notion that the rocks and the fossils say NO to evolution. As I will show here, the rocks and the fossils say YES to evolution!

The Main Geological and Paleontological Arguments of Scientific Creationism

The geological and paleontological arguments made by scientific creationists vary somewhat from author to author, and many rather long lists of presumed failings of the rock and fossil records have been published, but there is a group of recurrent arguments that I will discuss under four main headings.

1. *Evidence of catastrophe in the fossil record.* It is a common plea of the creationists that the geological record shows ample evidence of very sudden events, and they consider this anathema to the rather gradualistic, uniform processes of geology and evolution described in many basic textbooks. The examples presented by the creationists are legion, but emphasis is usually given to what are referred to as *fossil graveyards;* that is, those situations where there is evidence of sudden annihilation of populations of single species or of whole communities. The preservation of Pleistocene mammals in the tar pits at La Brea is one of many cases often cited. Emphasis is also given to *polystrate fossils.* This refers to situations where a single fossil specimen, such as an upright tree trunk, cuts across or is included in rocks covering a significant span of geologic time. The creationists point out, probably correctly in most cases, that the occurrences of a long tree trunk in life position suggests extremely rapid deposition of the surrounding rock. Otherwise, the tree trunk would have decayed and disappeared before it could be embedded. Fossil graveyards and polystrate fossils are combined to argue for the general principle that most fossils are the result of some unusual, short-lived event and do not represent a gradual or uniform process.

In a slightly different context, *ephemeral markings* are often cited as evidence of catastrophe. These include such sedimentary features as ripple marks, rain drop impressions, and mud cracks. Also included are footprints (of dinosaurs, for example) and a host of other biological markings. The argument is again made that these tracks and trails could not be preserved without some sort of unusual catastrophe.

2. *Relative dating based on fossils.* As is well known, geological dating normally takes two forms: *relative dating* developed empirically from the sequences of fossils and absolute dating based on a variety of techniques yielding an age in thousands or millions of years. The scientific creation-

ists have long criticized the system of relative dating based on fossils. One of their arguments is the claim that the basic reasoning is circular: a geologist identifies fossils as being of a certain age, they say, only because those fossils have been found only in rocks of that age. In addition, the creationists note that the entire column is never found in one simple stack and that the geologic column of the textbooks is actually a composite built up from small segments scattered around the world. The creationists argue that there is a large element of inference in the process of building up the chronology, and they have been quick to find fault with many of the details of the composite or standard column.

One also runs into the argument that the system of relative dating is circular because it assumes evolution; that is, fossils are placed in the sequence by their "stage of evolution," and the sequence itself is later used as evidence for evolution. As will be shown below, this particular point is a misunderstanding of the way geology works. Rarely, if ever, is the stage of evolution used as a means of placing a fossil in the geologic time scale.

3. *Absolute dating.* This is a complex subject and creationists' arguments deal primarily with alleged discordances or incompatibilities in dating methods using radioactive isotopes. The basic argument is that the number of inconsistencies in radiometric dating is great enough to disqualify the method.

4. *Disagreement between the fossil record and the predictions of Darwinian theory.* To the scientific creationist, the Darwinian theory of evolution predicts that we should find in the fossil record a continuous chain of evolutionary stages with ample intermediate or transitional forms between major groups. For authority, creationists quote Darwin, and indeed it is a simple matter to find in Darwin's writings the prediction of gradual evolution with intermediates strung out as beads on a string. The creationists then point to the rather sudden appearance in the fossil record of many new groups and the general lack of intermediates as evidence that Darwinian theory does not hold up. Also, in a slightly different vein, the creationists cite the so-called living fossils—those organisms that have shown little or no change through long periods of geologic time. It is not clear to me that Darwinian or neo-Darwinian theory predicts that evolutionary change must occur—so the living fossil argument may be a straw man.

Later in this essay I will consider the several arguments just presented in greater detail. But it is important, first, to understand what the creationists are *not* arguing, so that their real arguments can be properly addressed.

The Subtler Arguments of the Creationists

The biblical accounts of creation and of the history of the earth are often cited in the literature of the scientific creationists. Many creationists obviously believe that the Bible is correct in every detail. In fact, to become a member of the Creation Research Society one is obliged to subscribe to a statement of belief that includes agreeing that "the account of origins in Genesis is a factual presentation of simple historical truths." With this as background, the creationist could be content simply to present, and perhaps interpret, the biblical account and leave it at that, with no reference to observational data from natural history. But this is not the approach. Rather, they claim, the biblical account is used as a model or hypothesis, and its predictions are tested with data from geology, paleontology, and other fields.

Several lists of "predictions" of the creation model have been published (Gish 1978, pp. 50–51, for example). Testing these predictions often involves rather elaborate, and sometimes surprisingly conventional, research studies. A recent example is a reappraisal of the well-known limestone deposits of Silurian age at Thornton Quarry in Illinois (D'Armond 1980). These deposits are conventionally understood to be buried reefs, and extensive work has been done on them over many years. D'Armond attempts to argue that the deposits are simply the result of catastrophic flooding, and while I do not agree with his analysis or his conclusions, the study is clearly an attempt to use geologic data to support an aspect of the creation model.

Thus, while practitioners of scientific creationism firmly believe in the authority of the Bible, they do not attempt to rely on it as their sole authority. Rather, they appear to be searching for corroborative data from a wide range of sources. Theoretically, a creationist such as D'Armond could conclude that the creation model is not viable because of a lack of corroboration from geologic data. This is exceedingly unlikely for the committed creationist, but the literature of scientific creation does provide the interested layman with the opportunity to conclude that the biblical account is falsified by scientific data. In a real sense, creationists are putting the biblical account of creation on the line by claiming that it *should* be subjected to scientific testing.

It also comes as a great surprise to many people that contemporary scientific creationists claim to accept Darwinian natural selection and its modern genetic basis. That is, the creationists grant that populations of species are variable, that the variability is heritable, and that through natural selection evolution from one form to another takes place. Thus, classic cases of natural selection such as industrial melanism—the increased frequency of heavily as opposed to lightly pigmented peppered moths in industrial regions of Europe—do not bother the creationists. In fact, their textbooks often include exhaustive and sometimes even accurate treatments of the works of geneticists and population biologists on natural selection. The creationists do draw the line, however, at using natural selection to explain the origin of major groups (families and orders) and, of course, the origin of one species—*Homo sapiens*, as the following quotation illustrates:

> Creationists recognize that variation and mutation and natural selection are real processes but they feel that evolutionists are not justified scientifically in extrapolating from the essentially trivial cases of mutation and natural selection which can be observed to occur in the present world to the gigantic sequence of evolutionary changes which must have occurred in the past if the organic world is to be accounted for on this basis. . . . When all is said and done . . . examples of supposed present-day evolution that are commonly cited in textbooks are actually nothing but relatively minor variations within the originally-created kinds. . . . Essentially stability of the created kinds is postulated, though with a wide range of adaptive variety possible within the kinds. [Boardman, Koontz, Morris 1973, pp. 39–40]

It is thus the creationists' argument that the basic groups of organisms were created separately and that each created kind has undergone modification by perfectly conventional Darwinian means.

The creationists rely heavily on the idea that a single major flood was responsible for much of what we see in the geologic and paleontological records. This is inspired, of course, by the biblical Noachian Deluge. Still, the creationists do not insist that there be only one such flood, but rather they claim one large flood followed by an attenuated series of smaller floods. They argue that the preflood condition—an earth covered by crystalline rocks—lasted for some thousands of years and that this was followed by several months of flooding (Wysong 1976). The flooding is seen as producing the complex stratigraphy of sedimentary deposits that historical geologists and paleontologists deal with. Although there is some disagreement among creationists, the consensus is that there were a number of floods subsidiary to and following the Flood of Noah.

The Rocks and Fossils Say Yes!

In this section I will respond to the four main arguments of the scientific creationists presented earlier in this essay.

1. *Catastrophism.* The catastrophism argument is a straw man. In the nineteenth Century, the combination of Lyellian geology and Darwinian biology did promote a conventional wisdom that the earth and life evolved by very gradual processes moving at uniform rates. Many of the examples of catastrophism now being cited by the scientific creationists were well known but were either ignored or given very secondary importance in nineteenth-century geology and paleontology. A great deal has changed, however, and contemporary geologists and paleontologists now generally accept catastrophe as a "way of life," although they may avoid the word catastrophe. In fact, many geologists now see rare, short-lived events as being the principal contributors to geologic sequences. In many instances, an exposure of rock records a series of special events (storms, hurricanes, landslides, slumps, or volcanic eruptions) that produced large volumes of sediment but that represent only a fraction of the elapsed time covered by the total sequence. The periods of relative quiet contribute only a small part of the record. The days are almost gone when a geologist looks at such a sequence, measures its thickness, estimates the total amount of elapsed time, and then divides one by the other to compute the rate of deposition in centimeters per thousand years.

The question then is *not* whether catastrophes occurred (including large floods) but whether they were relatively few in number, with one large flood dominating geologic history. Assuming that our geologic time scale is reasonably accurate, geologists and paleontologists have identified many thousand separate catastrophic events, which means that any scenario based on catastrophism must include a very much larger number of small and large catastrophes than is allowed by the creationist model. Therefore, the general argument concerning catastrophism is a nonargument. Creationists claim that geology says that there should be no catastrophes. Creationists find some catastrophes and geologists find many—far more than are suggested by the creationist model. I suspect that the problem results from a basic misunderstanding of geology as it is now practiced. The misunderstanding has been caused in part by the geologists themselves: the nineteenth-century idea of uniformitarianism and gradualism still exists in popular treatments of geology, in some museum exhibits, and in lower level textbooks. It is even still taught in secondary school classrooms, and one can hardly blame the creationists for having the idea that the conventional wisdom in geology is still a noncatastrophic one.

2. *Relative time scales.* The charge that the construction of the geologic scale involves circularity has a certain amount of validity. It is true that we date fossils on the basis of our experience with the temporal distribution of the same fossils elsewhere. If one finds a totally new fossil on a roadside, it is impossible to place it in the geologic time scale because it is not in association with rocks or fossils of known age. Thus, the procedure is far from ideal, and the geologic ranges of fossils are constantly being revised (usually extended) as new occurrences of specimens belonging to the same species are found. In spite of this problem, the system does work! The best evidence for this is that the mineral and petroleum industries around the world depend upon the use of fossils in dating. If an oil company learns that petroleum is found in buried reefs of Silurian age, for example, its geologists search for reefs of Silurian age elsewhere. It has been shown over and over again that by following this strategy, more petroleum will be found than if drilling is done on a random basis. I think it quite unlikely that the major mineral and petroleum companies of the world could be fooled.

Another important element of this argument is that the use of fossils in geologic dating is in no way dependent upon biological theories of evolution. The best evidence for this is that the geologic column as we know it was quite fully developed by about 1815, nearly half a century before Darwin published *The Origin of Species.* In other words, the geologic chronology was developed on the basis of fossils before we had any Darwinian theory, and it was developed by people who subscribed largely to a creationist view of life. Thus, geologists using a creationist paradigm developed the geologic column, and only later was evolutionary theory added as a means of understanding or interpreting the sequence of fossils found in the rocks. It is in this context that I would note that fossils would work just as well in geologic chronology if they were only funny marks on rocks. Geochronology depends upon the existence of a virtually exceptionless sequence of distinctive objects in rocks; that sequence just happens to exist in the fossils.

The idea that geologists date a rock by the stage of evolution of its fossils is so deeply ingrained in creationist thought that it needs more discussion. In describing how the geologic column was developed, creationists have written that "the standard column was developed on the basis of the assumption of evolution. The fossils of 'early' ages are characterized by simplicity, of 'later' ages by complexity, because evolution must theoretically have proceeded generally in this manner" (Boardman, Koontz, Morris 1973, p. 33); and in a similar vein that "fossils are gathered from around the world . . . and assembled in a progressive order from simple to complex on a chart" (Wysong 1976, p. 353). Nothing could be farther from the truth.

I have already noted that the geologic column was constructed before Darwin, but there are other problems with the quotations just given. The method described by Boardman and Wysong would not work even if it were tried. To be sure, the oldest known fossils are of rather simple prokaryotic organisms and younger rocks contain more complex forms (multicellular eukaryotic organisms), but there is no recognizable trend toward increased complexity that is clear enough to use for dating purposes. This is in part because complexity is so difficult to measure: is an insect more or less complex than a starfish? This is an "apples and oranges" problem that defies a rigorous metric. Even where the fossil record of a coherent group of organisms can be traced for long periods of time, increasing complexity through time is elusive at best. (This is one of the interesting aspects of evolution: the process is not clearly directional.) Also, for the creationists' view of the way geologic dating works to be true, "simple" organisms would have to become extinct to make way for "complex" organisms so that the fossils in a given rock would give a clear signal. In fact, many primitive prokaryotes (bacteria, blue-green algae, etc.) are still living today—apparently quite happily. For this reason alone, stage of evolution could not be used to build a geologic time scale.

The whole problem is made more difficult by the fact that a surprising number of geologists with specialities other than paleontology share the same misconceptions. Wysong takes obvious pleasure in quoting W. M. Elsasser in the *Encyclopaedia Britannica* (1973) as saying, "The geological method presumes the existence in these periods of living beings of gradually increasing complexity" (1976, pp. 352–53). Professor Elsasser is an excellent geophysicist, but his expertise in fields distant from geophysics cannot be expected to be optimal. The creationists (and probably Professor Elsasser) come by their misunderstanding honestly, at least in part. Many teachers and textbook writers, especially in the late nineteenth and early twentieth centuries, have been so carried away by the elegance of the Darwinian model that they have ascribed powers to it that do not exist. It would be a fine thing if we could use some abstract estimate of stage of evolution to date rocks—but we cannot!

An interesting irony in this whole business is that the creationists accept as fact the mistaken notion that the geologic record shows a progression from simple to complex organisms. Faced with the problem of reconciling this presumed sequence with rapid deposition by the Flood, the creationists develop painful explanations of the sequence: large mammals floated to the surface of the Flood sea, complex (and therefore more mobile and intelligent) animals were able to escape to higher ground, and so on. The creationists have fit essentially false information into their model—something that would have been quite unnecessary had they read the geologic literature more carefully.

3. *Absolute dating.* The use of radioactive isotopes in geologic dating has many problems. The methods are inexact and contain many sources of error. In order for the system to work, the parent isotope must enter the rock in the absence of any of its daughter products, and the accumulation of daughter products must be contained in a closed system so that there is no leakage of daughter products out of the rock or migration of indistinguishable isotopes into the rock after it is formed. Furthermore, the half life of the radioactive isotope must be well known. The last assumption is apparently on firm ground, but the others are always subject to problems and errors. This means that a series of dates run on a single rock may produce quite different results, either because of leakage or contamination or because different isotopes record different events in the geologic history of the rock. Of all the methods, probably carbon 14 is the least dependable, and yet it is the most interesting to many people because it is applied to the most recent part of geologic

history. In spite of all the difficulties, however, radiometric methods do work well statistically; that is, there are enough concordant dates that the method is successful in dating rocks. One of the best pieces of evidence for this is the fact that the relative ages of rocks based on fossils correlate extremely well with the absolute ages of the same rocks based on radiometric methods. The correlation is excellent even though the two methods are as nearly independent as any two methods of measuring time could be.

The most significant finding of radiometric dating, of course, is that the earth is extremely old, perhaps 4.5 billion years old, and that life on earth is almost as old. This is in direct conflict with the ten thousand-year-old earth of scientific creationism. Although there could be some error in radiometric dating (and probably is), it is inconceivable, to me at least, that the error could be anything approaching the difference between billions of years and thousands of years.

4. *Darwinian predictions.* Darwin predicted that the fossil record should show a reasonably smooth continuum of ancestor-descendant pairs with a satisfactory number of intermediates between major groups. Darwin even went so far as to say that if this were not found in the fossil record, his general theory of evolution would be in serious jeopardy. Such smooth transitions were not found in Darwin's time, and he explained this in part on the basis of an incomplete geologic record and in part on the lack of study of that record. We are now more than a hundred years after Darwin and the situation is little changed. Since Darwin a tremendous expansion of paleontological knowledge has taken place, and we know much more about the fossil record than was known in his time, but the basic situation is not much different. We actually may have fewer examples of smooth transition than we had in Darwin's time because some of the old examples have turned out to be invalid when studied in more detail. To be sure, some new intermediate or transitional forms have been found, particularly among land vertebrates. But if Darwin were writing today, he would probably still have to cite a disturbing lack of missing links or transitional forms between the major groups of organisms.

How does the evolutionist explain the lack of intermediates? I see three principal areas of explanation, all of which probably operate to some degree. The first of these is a simple artifact of our taxonomic system of classification. The practicing paleontologist is obliged to place any newly found fossil in the Linnean system of taxonomy. Thus, if one finds a birdlike reptile or a reptilelike bird (such as *Archaeopteryx*), there is no procedure in the taxonomic system for labeling and classifying this as an intermediate between the two classes Aves and Reptilia. Rather, the practicing paleontologist must decide to place his fossil in one category or the other. The impossibility of officially recognizing transitional forms produces an artificial dichotomy between biologic groups. It is conventional to classify *Archaeopteryx* as a bird. I have no doubt, however, that if it were permissible under the rules of taxonomy to put *Archaeopteryx* in some sort of category intermediate between birds and reptiles that we would indeed do that. Thus, because of the nature of classification, there appear to be many fewer intermediates than probably exist.

In this context, it should be noted that creationists occasionally make the argument that the Darwinian model should predict a complete absence of distinct kinds of organisms.

> If all organisms have actually descended by evolution from common ancestors, it seems inexplicable that there should be any distinct categories of organisms at all. One would certainly expect that nature would instead exhibit a continual series of organisms, with each grading into the other so imperceptibly that any kind of classification system would be impossible.
>
> [Boardman, Koontz, Morris 1973, p. 68]

This, unfortunately, shows a lack of understanding of the separation of genetic systems through reproductive isolation. There is little or no gene flow between species because they do not normally interbreed. Thus, each species is able to evolve on a course independent of all others, and there is no opportunity for blending once speciation has taken place. Given time, and perhaps subsequent speciation events, organisms become distinct. By the same reasoning, major groups such as molluscs and arthropods become increasingly distinct and separated by anatomical gaps. Thus, the presence of distinct kinds of organisms (especially when viewed at an instant in time) is a reasonable prediction of the evolutionary model. Because the creationist model also predicts distinct kinds

(Gish 1978), their mere presence cannot be a basis for argument between the two viewpoints. The only argument is whether the historical record of fossils should show more transitions between the distinct kinds than it does.

A second line of explanation for the underrepresentation of intermediates is the same one that Darwin used, namely that the fossil record is incomplete. We have as fossils a tiny fraction of the species that have existed. There are many ways of documenting this, but one is simply to look at the comparative numbers of extinct and living species. There are something like 2 million species known to be living today. We know that the average duration of a species is short relative to the total span of geologic time. Therefore, there must have been turnover in the species composition of the earth many times since the beginning of the fossil record. If we had even reasonably good fossil preservation, the number of known fossil species should thus be some large multiple of the number of species living today. Yet only about a quarter of a million fossil species have been found. This can only lead to the conclusion that the odds against fossilization are so high that we are seeing just a tiny fragment of past life. Also, along the general idea of catastrophism, the fossils that we do see depend largely upon occasional or unusual physical and biological events, and therefore the record is not a uniform or random sampling of life of the past. Under these circumstances, finding transitional forms (or any other particular form) is unlikely, and it is thus not surprising that our record appears to be quite uneven and jerky. In addition, most major groups of organisms originated quite early in the geological record, in that part of it that is especially poorly documented and where intermediate forms would be even less likely to be found. In this context, it is not surprising that our best intermediate or transitional forms are among land vertebrates, which evolved rather late in geologic time.

A third general explanation for the relative lack of intermediates is that transitional forms constitute very short intervals of geologic time if, as many evolutionary theorists now believe, the change from one major type to another occurs rather rapidly (the punctuated equilibrium model of Eldredge and Gould 1972). This simply lessens the probability of finding intermediates.

With these considerations in mind, one must argue that the fossil record is compatible with the predictions of evolutionary theory.

Creationism and the Integrity of Science

It is often argued that the creationists have allegiance to a single ideology (the Bible) and are thus not free enough intellectually to consider questions of origins in a scientifically acceptable manner. There is no question that there is a strong correspondence between support of the creationist idea and commitment to a single religious view. Increasingly, however, people without strong religious commitment are being drawn into and are expressing some acceptance of the arguments made by the scientific creationists. Therefore, control by an ideology may represent an argument in some quarters but certainly not in all. Furthermore, I think it can be argued that whether a body of reasoning is scientific or not should be decided independently of the question of whether the adherents are committed to one ideology or another.

In my view, a few of the arguments used by the creationists are "scientific" in the sense that they use the basic methods of testing hypotheses normally considered to be scientific. This does not mean, of course, that the conclusions are correct. Bad science may be difficult to distinguish *methodologically* from good science.

In spite of my claim that at least some of the scientific creationists are behaving, superficially at least, scientifically, they do display major problems in execution. Some of these are simple errors of fact or of understanding of the ways in which evolutionary biology and paleontology operate. For example, it was noted earlier that scientific creationists argue that it is invalid for geologists and paleontologists to determine the age of fossils on the basis of some presumed level or grade of evolution. This is a clear misunderstanding of the way geology and paleontology operate. The development of the relative time scale for the fossiliferous part of the geologic column is purely empirical. The fossils could be any sort of funny marks on rocks, unrelated to biological entities, as long as they are nonrandomly distributed in time.

Other errors of fact are illustrated by the often referred to simultaneous occurrence of dinosaur and human footprints in Cretaceous limestones in Texas. This is one of many instances where lack of paleontological training and lack of experience with fossilization have led the creationists astray. In this particular case, the dinosaur footprints are real but the "human footprints" are not: that is the judgment of vertebrate paleontologists who have worked extensively with fossil trackways of terrestrial tetrapods. So, although some scientific creationists have done a rather remarkable job of absorbing a complex discipline, errors of fact and of understanding have crept into their work.

A more serious deficiency in the scientific method used by the creationists is their repeated insistence on experimental evidence and their insistence that there be no exceptions. The creationists are fond of claiming that for something to be scientifically demonstrable, it must be amenable to proof by experiment and it must be without exceptions. These requirements are probably valid in certain areas of science, particularly in parts of physics and chemistry and in certain areas of engineering. What the creationists seem to miss is the fact that geology and paleontology are historical sciences, and therefore experimental testing of predictions is difficult, and that these sciences rely largely on statistical inference—that is, on the building of a general case that accepts exceptions as tolerable, especially when there is a highly plausible explanation of those exceptions. In this context, the kind of inference made by geologists and paleontologists is not unlike that made in clinical medicine where both diagnosis and treatment are inexact and individual decisions may depend upon assessment of probabilities and predictions that may, in some cases, turn out to be incorrect. The "batting average" is high enough in most areas of medicine to justify these fields as being not only scientific but well worth the effort. Clinical medicine is often an historical science in that inferences and generalizations are based on past events that were totally unplanned.

Let me give an example of an accepted exception in a geologic context. The basic geologic column was developed on the assumption of the so-called Law of Superposition. This law simply says that younger rocks are deposited on top of older rocks and therefore that if one finds a sequence of rocks, the youngest are at the top and the oldest are at the bottom. This is not a very profound law, but it has been extremely useful and was vitally important in the development of the geologic time scale. Not uncommonly, however, demonstrably young rocks are found *beneath* older rocks. Often the reason for this reversal of the expected sequence is clearly the result of movement of the rocks by tectonic forces after deposition, specifically by thrust faulting (when one set of rocks is literally thrust up over a younger set of rocks long after the original sequence was deposited). Under ideal conditions, one can find clear evidence of thrust faulting and can even identify the surface along which the movement took place. In such situations, the reversal of the order is not a meaningful exception to the Law of Superposition.

With many well documented cases of thrust faulting in hand, the geologist feels confident in interpreting a reversed sequence as the result of faulting even when actual evidence of the fault cannot be found in the particular case. This practice is dangerous, of course. The interpretation of such discordant sequences could be in error, but the geologist is comfortable with the reasoning because the number of unexplained exceptions to simple superposition is *very* small compared with the number of situations where the expected sequence is found or where a clear explanation for the disturbed sequence is available. It is in this sense that the geologist is making a statistical argument when he interprets a reversed sequence by recourse to thrusting.

The creationists appear content to cite one or a half dozen unexplained cases of reversal to disqualify the whole system of geologic chronology. Actually, what they should be trying to do is to build up a statistical argument wherein the number of unexplained exceptions is so large as to jeopardize the entire reasoning. They have not been able to do this, and I suspect they have not tried because of their basic thesis that a theory or law can be brought down by a small number of exceptions.

In summary, my feeling is that to the extent that the better students of scientific creationism are using scientific methods, they are not using them well.

Could the Evolutionists Be Wrong?

It would be folly for evolutionists to claim that they have a complete and accurate understanding of the history of life and of the processes that produced that history. Too many major paradigms in science have been overturned for any statement of such absolute confidence to be wise. We should consider alternatives and we should consider the possibility that we might be wrong in at least some parts of the basic framework of evolutionary thinking. And this consideration of alternatives is, in fact, going on in the 1980s with challenges from within evolutionary biology itself to the neo-Darwinian model as it is applied to macroevolution (Lewin 1980).

There are some basic aspects of evolution, however, that are so close to being simple observation and measurement that evolutionists can claim to be right. In particular, geologic dating (both relative and absolute) is on extremely firm ground. To challenge the basic chronology of life forms would be like claiming that the sun is only ten thousand miles from the earth or that the earth is flat. In effect, we can "see" the geologic time scale. If organic evolution is defined as change in the biological makeup of life on earth over time then we certainly do have evolution and can "see" the fossil record of that process.

Deducing the mechanisms of evolution is quite a different matter. We are confident that the process of natural selection works at the population level, and there is no argument about this between the evolutionists and the creationists. But we are not sure whether we can extrapolate this process of *microevolution* to explain the larger events of *macroevolution.* Even if it turns out that the classical Darwinian model does not explain some aspects of evolution, we will not be obliged to shift to a creation model. The literature of evolutionary biology and paleobiology contains a host of alternative *biological* models, and these are being evaluated and tested in many separate research projects. Thus, the scientific creationists are totally wrong in their so-called two model approach—the claim that if the Darwinian model is discredited, the only alternative is the creation model.

REFERENCES CITED

Boardman, William, Koontz, Robert F., and Morris, Henry. 1973. *Science and creation.* San Diego: Creation-Science Research Center.

D'Armond, D. B. 1980. Thornton quarry deposits: a fossil coral reef or a catastrophic flood deposit? A preliminary study. *Creation Research Society Quarterly,* 17:88–105.

Eldredge, Niles and Gould, Stephen J. 1972. Punctuated equilibria: an alternative to phyletic gradualism. In *Models in paleobiology,* ed. T. J. M. Schopf, pp. 82–115. San Francisco: Freeman, Cooper and Co.

Gish, Duane T. 1978. *Evolution? The fossils say no!* San Diego: Creation-Life Pub.

Lewin, R. 1980. Evolutionary theory under fire. *Science* 210:883–87.

Wysong, R. L. 1976. *The creation-evolution controversy.* Midland, Mich.: Inquiry Press.

10

Systematics, Comparative Biology, and the Case against Creationism

BY JOEL CRACRAFT

Systematics is the study of organic diversity through space and time. The emphasis of systematics is on the theory and methodology that is applied to the historical analysis of diversity rather than on the atemporal aspects of diversity that more properly might belong to the field of ecology.

Viewed in this way, systematics investigates problems of species, primarily the ways in which they are thought to originate, their genealogical relationships with other species, and the historical aspects of their distribution in space and time. To investigate these questions systematists use the methods of comparative biology, including biogeography. Systematics plays a central role in modern evolutionary biology and contributed significantly—in the guise of comparative anatomy and biogeography—to the formulation of Darwin's (1859) ideas on evolution. Conceptually, however, the field of systematics has made significant, if not revolutionary, advances over the last two decades, and it is becoming increasingly apparent that systematics is the area of biology that defines the *pattern* of organic change through space and time and, consequently, specifies that body of knowledge that theories of evolutionary *process* must be capable of explaining. Systematics, therefore, provides an important focus for many of the recent controversies arising from the creationist challenges to evolutionary biology.

Evolutionary Species versus "Created Kinds"

The notion that we can categorize, or classify, organisms into groups on the basis of shared similarity is ancient indeed. So too is the realization that there exist groups of individual organisms that are defined in terms of their ability to interbreed with one another and thereby produce like kinds, but that at the same time lack the ability to interbreed freely with other such groups (although accidental interbreeding might be observed sometimes). These interbreeding groups are the entities biologists call species, and the concept of reproductive cohesion has been used historically as a criterion for defining their unity or individuality, regardless of whether the investigator believed in a supernatural creation of species or in their evolutionary origin.

Accompanying the reproductive criterion has been the recognition that organisms show variability not only within species but among them as well. In fact, morphological discontinuity among clusters of like individual organisms has been the single most important basis for recognizing species limits, especially because information about interbreeding is known for only a very small percentage of the species currently recognized.

For a long time, biologists have known that morphological variation within a reproductively defined species is often greater than that existing between two *sympatric* (i.e., living in the same area) species that are virtually identical morphologically but that are themselves reproductively isolated. In cases such as this, the phenotypes of the two sympatric species usually exhibit slight differences in morphological or behavioral characteristics that have a significant influence on their reproductive isolation (see Mayr 1963, chapters 3–5). The point to be made here is that the presence of reproductive isolation is not necessarily related to the degree of phenotypic difference, and at least in sympatric taxa it is their reproductive discontinuity that is of significance when naming them as distinct species.

When distinct but obviously similar forms occur in different geographic areas (a distribution pattern termed *allopatric*), it is not possible to apply a criterion of interbreeding. Traditionally, systematists have relied on the degree of difference of these isolates, as compared to the differences shown by sympatric pairs that are close relatives, to establish whether they are separate species or whether they are best treated as subspecies (geographic races) of a single species. In these cases the most important observation of scientific interest is the existence of differentiated taxonomic units; whether these units are treated as species or subspecies is less important.

For over one hundred years, biologists have expended intense effort in describing the different kinds of organisms, in discovering how their phenotypic variation is expressed geographically, and in investigating their reproductive relationships to similar taxa. This vast biological literature represents a substantial body of knowledge on the subject, and thus the superficial and illiterate treatments of this knowledge in the creationist literature calls into question the creationists' claim of scientific objectivity and competence. Creationist views of the importance and recognition of the "kinds" of organisms derive solely from a literal acceptance of the Bible rather than from the application of a scientific methodology to data on variation, distribution, interbreeding, and ecology. Consequently, it should come as no surprise that creationists ignore most of the conventional procedures of systematics. The depth of their scientific acumen is illustrated by a particularly elementary example: they seldom even refer to taxa by their scientific names, preferring instead to adopt scientifically imprecise names such as dog, cat, bat, horse, and so on. To creationists, the species of modern biology should be replaced by the "created" or "basic" kind:

> A basic animal or plant kind would include all animals or plants which were truly derived [presumably by special creation] from a single stock. In present-day terms, it would be said that they have shared a common gene pool. All humans, for example, are within a single basic kind, *Homo sapiens*. In this case, the basic kind is a single species. . . . In other cases, the basic kind may be at the genus level.
>
> [Gish, 1979, p. 34]

A vague notion of genetic compatibility seems common to most creationists' concept of kinds:

> The oft-repeated statement, however, that God's creatures brought forth progeny 'after their kind' would strongly indicate that plants and animals which can interbreed and produce offspring would be the same 'kind.' A corollary conclusion would then be that production of offspring from matings between two different kinds would be impossible. [Siegler 1978, p. 37]

In this quote Siegler seemingly identifies fertile offspring as the criterion for defining kinds, but on the same page he contradicts this and restricts "kinds" to those organisms that can effect fertilization, regardless of whether the zygote remains viable. He calls these kinds "baramins" (from the Hebrew *bara,* created, and *min,* kind, after Marsh 1947), and notes that they may apply to any portion of the conventional taxonomic hierarchy. Thus, *Homo sapiens* is a baramin at the species level, and at the upper extreme Siegler suggests the entire family of waterfowl (Anatidae) is a baramin or single "created kind."

Siegler's apparent confusion over just how a "created kind" should be recognized (fully viable offspring or merely ability to fertilize) is paralleled in other creationist writings:

> We cannot always be sure, however, what constitutes a separate kind. The division into kinds is easier the more divergence observed. . . . Among the vertebrates, the fishes, amphibians, reptiles, birds, and mammals are obviously different basic kinds. . . . Within the mammalian class, duckbilled platypuses, opposums, bats, hedgehogs, rats, rabbits, dogs, cats, lemurs, monkeys, apes, and men are easily assignable to different basic kinds. Among the apes, the gibbons, orangutans, chimpanzees, and gorillas would each be included in a different basic kind. [Gish 1979, pp. 36–37]

This is a very revealing passage, one that would seem to raise a serious problem for the creationists and especially for Gish. If we accept for the moment the creationist position—that the various basic kinds were created *de novo* apart from one another and that basic kinds can generally be recognized by some criterion of interfertility (e.g., Gish 1979, p. 36)—then one must conclude that Gish, a leading creationist, does not understand what a "created kind" really is. This conclusion follows logically from his own statement: an organism cannot be a "created kind" in and of itself (the gibbon, the orangutan, the chimpanzee, and the gorilla in the above quote) and at the same time be an element of another, more inclusive "created kind" (e.g., "apes," or mammals). Given the notion of "created kinds" accepted by most creationists (Whitcomb and Morris 1961, pp. 66–67; Morris 1974, p. 180; Wysong 1976, p. 58), Gish's position is logically indefensible. Moreover, one could scarcely maintain "fishes," mammals, bats, and so forth as single "created kinds" on the basis of *any* criterion of fertility. If, on the other hand, Gish wishes to maintain the accuracy of his statement, then the concept of "created kind" becomes vacuous: essentially, it is anything a creationist says it is. Naturally, most creationists probably would not want to adopt this position, and thus we might expect most creationists would disagree with Gish's interpretation of "created kinds." It is curious, therefore, that none of them have sought to challenge Gish's interpretation and to discuss the question of "created kinds" in a critical manner. Then again, because the concept of "created kinds" is a manifestation of a fundamentalist religious belief rather than of a scientific attempt to understand nature, why should we expect the creationists to adopt a critical attitude?

If the concept of "created kinds" seems inapplicable because of creationists' inability to clearly express what it means, the lack of any rigorous methodology for investigating these supposed entities underscores the weaknesses in their entire approach to science: they misunderstand or distort modern thinking within evolutionary biology and, once the latter is summarily rejected, replace it with a naïvely conceived substitution. Let's examine a case history.

Lammerts (1970) studied specimens of the Galapagos (or Darwin's) finches in the collections of the California Academy of Sciences. He compared coloration among the species and measured body (something a professional ornithologist would never do because of effects of preservation) and bill size. He then observed that "if one were to remove all the species labels and arrange the Darwin's finches from the largest to the smallest in body and bill size, complete intergradation would be found" (1970, pp. 360–61), and he goes on to argue that all these species are based merely on minute differences in size and shape and that, in reality, they should constitute only a single species, in this case probably one "created kind." Lammerts's conclusion has been cited by other creationists (e.g., Gish 1979, p. 36) as an example of how the species of evolutionary biology can be translated into "created kinds."

In his zeal to demonstrate how biologists have misunderstood the species limits within Darwin's finches, Lammerts exemplifies the creationists' inability to deal critically with scientific data without distorting it for their own ends. He conveniently does not mention that many of these species occur sympatrically with each other on one or more islands and fail to interbreed: they are reproductively discontinuous. Lack (1947), who is cited by Lammerts, makes specific note that hybridization among the species of Darwin's finches is rare, if it is present at all.

By all criteria that creationists themselves recommend, each species of Darwin's finch would have to be called a "created kind." This leads one to wonder why they would adopt the position that there is only one species of finch, based as it is on blatantly inaccurate biological information. I can suggest a number of obvious reasons, none of which creationists are very candid about in print:

1. Creationists would be uncomfortable if they had to admit that the species concept of evolutionary biologists, based as it frequently is on morphological criteria, actually does mirror the existence of reproductively discontinuous units. In so doing they would have to accept the reality of many more "created kinds" than they would prefer (one of their reasons for minimizing the number of kinds is so that they will all fit on the Ark!).

2. Creationists would then have to admit that species—reproductively distinct kinds—can evolve by isolation and subsequent differentiation. Creationists readily admit to evolutionary change *within* reproductively distinct "created kinds," but they deny the origin of these kinds from one another.

Unfortunately for them, creationists have used the Darwin's finches as an example of evolutionary change within what they assumed to be a single "created kind" (Lammerts 1970, p. 361; Gish 1979, p. 36). But by their own definition, these finches cannot represent a single "created kind"—each species is reproductively isolated. It follows, therefore, that creationists have essentially admitted—through their own error of not understanding the biological data—that new species can arise by isolation and subsequent differentiation. The only logical escape for the creationists would be to admit openly that they cannot define "created kinds" and cannot provide a methodology to identify them. But they do not dare do this, for it would repudiate their claims that "creationism" can be studied scientifically (see below).

3. Generally speaking, creationists are uncomfortable if their "created kinds" are not markedly distinct morphologically. The more similar their "created kinds" are to one another, the more difficult it is to deny the reality of evolution. It should be apparent that small changes accumulated during speciation, as in the Darwin's finches, when extrapolated through geological time provide a plausible basis for apparent large scale differences among groups of organisms. It is no doubt apparent to the creationists as well—hence their reluctance to conceive of "created kinds" narrowly.

4. If creationists admitted that small-scale changes can occur *between species* by evolutionary means as in the Darwin's finches—and I have argued that creationists effectively do this—then they would have problems in explaining the marked differences among the differentiated populations of humans. If new species of finches can originate by naturalistic processes, why not *Homo sapiens?* And the differences between *Homo sapiens* and some fossil hominoids are uncomfortably small in comparison to the variation among present-day human populations.

In summary, the concept of "created kinds" is essential to the acceptance of creationism. The "created kind" is the unit of a creation event just as the species is the unit of evolutionary change. Consequently, if the concept of "created kind" cannot be defined so that it can be used to interpret and investigate nature, then it is of little or no importance for the growth of knowledge. If it is not important for the growth of knowledge, creationists cannot maintain the charade that "creationism" provides a basis for the scientific investigation of nature. I have argued in this section that the creationist concept of "created kind" cannot be defined objectively without leading to the conclusion that one "created kind" can give rise to another by naturalistic means, which is nothing more than evolution as has been defined for centuries. Indeed, I have argued that creationists, in their confusion, have admitted to the evolution of "created kinds." They will, of course, vehemently deny this and will tenaciously cling to the concept of "created kinds," for it is, after all, only an article of religious faith, and of a narrow fundamentalism at that.

Biological Comparison: A Natural Hierarchy or Analogical Similarity?

Human beings communicate with language, and language is a classification system, with words generally signifying group membership of objects or concepts. The basis of any classification system is similarity. Long before the concept of evolution was accepted by the biological community, natural historians were making classifications of organisms and were attempting to identify those groups that could be called "natural" (for a recent history of these attempts, see Nelson and Platnick 1981). In preevolutionary times "natural" usually was interpreted to mean those groups assumed to be the product of a "creation event" and that evidenced a "divine plan." After the rise of an evolutionary viewpoint, natural groups were those thought to have descended from a common ancestor. In both cases some aspect of similarity was used to define the content of these natural groups.

For several centuries now, countless biologists have been discovering new forms of life—both living and fossil—and as a result of extensive comparison of their intrinsic attributes (morphological, physiological, biochemical, and behavioral), they have come to the realization that nature is hierarchical. This means that the groups of organisms (taxa) themselves can be arranged hierarchically on the basis of a repeatable, and therefore nonrandom, internested set of similarities. Consider the simple example of figure 1 (modified from Eldredge and Cracraft 1980, p. 25). If we make comparisons between the five organisms of figure 1, we find that they exhibit similarities in characters of embryonic and adult morphology and that these similarities can be used to construct a hierarchical arrangement of the organisms. This hierarchical arrangement of similarity is called a *cladogram* and is usually expressed as a branching diagram (fig. 1d), but it can also be expressed as a classification, much like the conventional Linnaean classification scheme used throughout biology (fig. 1e). A cladogram can be viewed as a scientific hypothesis, because it is a statement about the general pattern of similarity to be observed among these five groups. There is a prediction that further similarities will be congruent with the hierarchical arrangement of the cladogram (see Eldredge and Cracraft 1980; Nelson and Platnick 1981; and Wiley 1981, for modern treatments of this material). And a prediction of this kind can be tested empirically.

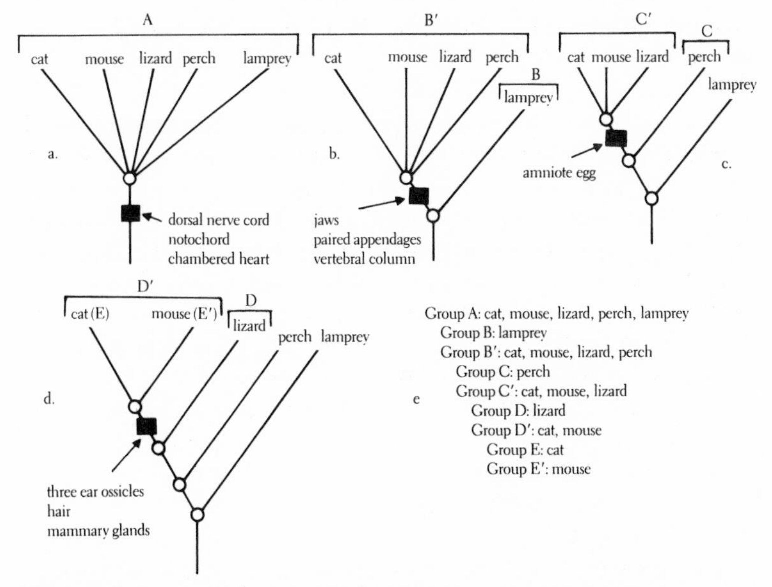

Figure 1. Similarities (characters) among organisms are shared in a pattern that is hierarchical in structure

These shared similarities, interpretable by biologists as evolutionary novelties (black squares of the figure), can be used to cluster the five organisms into groups and subgroups whose similarity relationships can be expressed either as branching diagrams (a-d) or as a classification (e).

In the past, branching diagrams of this sort have often been taken to express the pattern of general overall similarity, but a German entomologist named Willi Hennig showed that this was not the case. Hennig (1966) noted that similarities are of two basic kinds, nonhomologous (convergent) and homologous, and that the latter may be either primitive or derived. Hennig's chief insight was to demonstrate that branching diagrams should be based on derived similarity alone and that primitive and nonhomologous similarities cannot lead to the discovery of a natural hierarchy.

In evolutionary terms, derived similarities (or characters) are seen to be evolutionary novelties—for example, the feathers of birds or the hair of mammals. Both these characters are interpreted by biologists to be derivative of scales, such as those found in reptiles, and although feathers and hair can be used to define the two groups mentioned, *within* birds and mammals, feathers and hair are each seen to be primitive and thus cannot unite subgroups of birds or mammals.

To most comparative biologists, the concept of primitive and derived characters has evolutionary connotations, but it need not be interpreted in this way only.

Just as a broad comparison of adult morphologies can lead to a hypothesis about a primitive to derived sequence (scale to feathers, for example), which then can be employed to construct a hierarchy of groups within groups, so too can embryological differentiation lead to a hypothesis concerning whether characters are more general or widespread (e.g., generalized vertebrate egg) or whether they are less general or restricted in their distribution (amniote egg). Thus, embryological transformations can yield hypotheses about taxic hierarchies—they have been used for that purpose since the last century—without demanding an assumption of evolution (this is not to say, however, that an evolutionary interpretation cannot be applied).

The literature on the methodology of comparative biology (systematics) is large and relates in detail how comparative data can be used to construct branching diagrams; three recent summaries of these methods are Eldredge and Cracraft (1980), Nelson and Platnick (1981), and Wiley (1981). The remainder of this section will discuss how creationists have viewed the problem of similarity and, most importantly, will argue that the *hierarchical pattern* produced by the shared similarities observed among organisms is predicted by a hypothesis of evolutionary descent with modification but *not* by an assumption of special creation.

One explicit assumption of a creationist world view is that all the different "kinds" of organisms originated more or less at once, without any temporal connections of genealogy between them. This scenario is precisely the reason why the concept of "kinds" is so important for the creationist argument: within "created kinds" creationists admit to genealogical descent and modification, but between kinds they do not. If one accepts this interpretation of the natural world, and virtually all those claiming to be "scientific creationists" do, then at least one prediction follows logically from its underlying assumptions:

Creationist Prediction 1: *The similarities observed among organisms cannot be shared so as to produce a hierarchical pattern of groups within groups.*

A corollary prediction, discussed in virtually all creationists writings, is:

> Creationist Prediction 2: *All morphological similarities shared between separate "created kinds" will exhibit strong correspondences in functions and biological roles that are tightly correlated with parallel ways of life.*

First, let's examine the basis for these predictions. As noted earlier, creationists are predisposed toward the recognition of "created kinds" that are separated from each other by well-defined morphological gaps. But, as is obvious, the gaps are not complete—that is, distinctly different kinds of organisms share characters with one another. The creationists have erected an explanation to account for this (their explanation is the second prediction: that these similarities are a manifestation of functional design; this will be discussed below). Although creationists recognize the existence of shared similarities among disparate taxa and ascribe them to similarities in functional design, they also seem to realize that the assumption of creation would not lead one to expect a pattern of shared similarity that is consistently hierarchical in structure. On the other hand, modern comparative anatomy texts, with their admittedly evolutionary background, take note of this hierarchical pattern of similarity, and indeed it is a major prediction of the hypothesis that life evolved. In fact, such a pattern is less a prediction than it is a logical consequence of the evolutionary hypothesis; thus, it is not surprising that creationists have never claimed it as a prediction of their own world view. What is surprising, however, is that to my knowledge they have never stated explicitly the converse: that the *absence* of this pattern is a prediction of the creationist scenario. Among the creationists, Parker (1980) perhaps comes the closest to rejecting a hierarchical pattern. As an alternative, he suggests there is a "mosaic" pattern to the similarities among organisms. This is not a novel proposal, and through the history of systematics various workers have produced branching diagrams purporting to depict mosaic patterns of similarity (see the historical summary in Nelson and Platnick 1981). Mosaic representations of similarity relationships have been abandoned by the vast majority of biologists, and were by most of those who worked in preevolutionary times, for a very simple reason: hierarchical patterns convey more information and are less ambiguous about the group membership of those taxa, which to most biologists appear to have objective "reality." In a nutshell, hierarchical patterns tell us more about nature than do mosaic patterns. Let's consider a simple example. As is well known, whales share many characters with terrestrial mammals: hair, mammary glands, a single bone in the lower jaw, etc. It is also possible to find some similarities between whales and, say, sharks: aquatic habits, fusiform bodies, and vertebral columns. If one treats this list of characters equally, then a mosaic pattern of similarity can be recognized. But one does not have to be an evolutionist to know that these characters are not all equal: hair, mammary glands, and a single bone in the lower jaw seem to be similarities shared by all animals we call mammals; a fusiform body seems to be a similarity shared by many animals that locomote rapidly through water; and a vertebral column seems to be a similarity shared by all animals we call vertebrates. Presumably Parker realizes that there exists a group that we call mammals—remember that his colleague Gish called mammals a "created kind"!—and that whales are properly members of that group. And presumably Parker realizes that there is a still larger group called the vertebrates and that mammals are only one subgroup of them. If so, Parker and other creationists would have to admit to the logical conclusion that some similarities carry more significance than others for forming groups and that, as a result of these defining characters, these groups (such as mammals) have some objective reality. Barring a suspension of logical reasoning on their part, even creationists would then have to conclude that a hierarchical pattern is manifestly the most appropriate way of representing similarity relationships among organisms.

It would, of course, be possible for creationists to appeal to supernatural intervention to account for the existence of a hierarchical pattern of similarity—God might have, the creationists could argue, designed the world in just this way. On the other hand, it is unlikely that the creationists would want to take this form of argumentation, for two reasons: (1) the absence of this pattern would seem to be a logical outcome of a "creation event" that produced each "created kind" more or less simultaneously, which means that the creationists would then have to admit that their argument makes God appear illogical (Why would God create a pattern that appears to be the result of evolution?); and (2) the creationists ask—through their literature, public speaking, and legal actions—that we judge their position vis-à-vis evolution on purely scientific grounds, which means that an appeal to the supernatural would seriously question the basis of "scientific creationism."

The second prediction of the creationists is concerned with the problem of functional design. Thus, speaking of the shared similarities seen among different "created kinds," creationists offer the following explanation:

> Creationists, on the other hand, interpret the same similarities as evidence of common creative planning and design. The evolutionist has to assume all such characteristics have developed by chance mutations and natural selection. Creationists explain them as structures designed by the Creator for specific purposes, so that when similar purposes were involved, similar structures were created.
>
> [Morris 1974, pp. 69–70]

Assuming that both creationists and evolutionists would agree that it is not possible to investigate scientifically the cause of a particular design being the result of some "creative force," how are we to evaluate the creationists' explanation of similarity? Empirically, what the creationists predict is that all similarities among "created kinds" are designs for similar ways of life. Can this prediction be tested and is it substantiated upon examination of cases in nature?

Personally, I doubt whether the creationists' explanation of similarity —like much of the "adaptive" storytelling prevalent in evolutionary biology (Gould and Lewontin 1979; Cracraft 1981)—is rigorously testable. There are too many possibilities for differences of opinion over what should be identified as a similarity in morphology, a "purpose" (a function or a biological role), or a "way of life." Thus, presented with an instance of falsification, a creationist might argue that the organisms being compared are not really similar. For example, one might observe that a very long, fusiform body is a similarity shared by many invertebrates, various diverse groups of teleost fishes, snakes, and some amphibians. No reasonable person would accept an argument that this similarity was "designed" in order that they might inhabit a similar way of life, and thus this similarity would seem to falsify the creationist prediction. Nevertheless, interminable arguments could be raised that upon detailed observation these bodies are not actually "similar" or perhaps that the "purpose" of this similarity is actually the same for all (i.e., "locomotion") and that therefore we see a common design.

What can be concluded regarding creationism and the problem of design? The main point would seem to be that creationists should not invoke common design because not only are statements about similarity and functional design frequently too subjective and difficult to evaluate empirically, but they also are consistent with evolution as well. Evolutionists have traditionally interpreted morphological similarity as functional design for similar environmental conditions, but creationists are wrong when they assert that evolutionists must assume all similarities were produced "by chance mutation and natural selection." Such a claim merely reflects a superficial understanding of contemporary evolutionary biology and a narrow interpretation of evolutionary mechanisms. The concept of change being due to "chance" is philosophically and psychologically offensive to a creationist—it conjures up a world lacking purpose, direction, or design. But evolutionary change occurs not by "chance," if that word is taken to mean "at random," because the probability of evolutionary change in phenotype is not equal in all directions. The adult phenotype is the result of a highly regulated developmental (ontogenetic) history in which the phenotype is influenced not only by direct genetic controls over elaboration of biochemical products and their expression in developmental pathways, but also by epigenetic (environmental) factors modifying those pathways (Løvtrup 1974; Alberch 1980).

The development of organisms is thus canalized, or constrained; consequently, changes in the underlying genetic control or in environmental factors having an influence on ontogeny do not produce a random ("chance") array of phenotypic responses but rather a very narrow spectrum of possible alterations. In this way, then, much of evolutionary change can be viewed as being "directed" by developmental canalization, the exact direction being determined by a host of genetic and epigenetic factors.

Our increasing knowledge of developmental biology, and epigenetics in particular, is contributing to the view that developmental programs of different organisms may respond similarly if they are exposed to comparable environmental stimuli, even when those programs have differences in their genetic backgrounds (Løvtrup 1974; Alberch 1980). Thus, unlike the simplistic characterization of evolution proposed by the creationists (and unfortunately, by some evolutionists) in which natural selection is envisioned as the primary, if not only, mechanism of directional change, modern evolutionary biologists are realizing that the magnitude and directionality of phenotypic change is primarily a problem of developmental genetics.

Returning to the main theme of this section, there is abundant evidence in the systematic literature for the existence of a natural hierarchy. This hierarchy is discoverable through the comparative analysis of derived similarities, and by far the best explanation for this hierarchy is that it represents the pattern of the phylogenetic history of life. It must be stressed that conjectures about the phylogenetic history of any particular group are scientific hypotheses that are subject to critical evaluation. Indeed, there is often much debate within the systematic community over which phylogenetic hypothesis best explains the available comparative data. Nevertheless, the relationships of many groups are well understood, leading to the conclusion that the existence of a natural hierarchy is as well corroborated as any hypothesis in biology.

Where does this leave the creationists? I suggest that the existence of a natural hierarchy refutes the creationist world view. To my knowledge they have never denied the existence of a hierarchy of similarity, nor have they provided a credible, scientific explanation as an alternative to descent with modification. I am unaware of a single paper in the scientific literature in which a creationist has attempted to refute the hypothesis of a natural hierarchy. Furthermore, inasmuch as systematic biology is the very cornerstone of evolutionary analysis (Eldredge and Cracraft 1980; Wiley 1981), it is significant that apparently no creationist is a professional systematist publishing in the scientific literature (to criticize the ideas in a field as complex as systematics, it is necessary to have a rather sophisticated understanding of those ideas).

The Natural Hierarchy and Intermediate Taxa

Creationists repeatedly assert that there are no taxa morphologically intermediate between major groups, and they use this claim as a cornerstone of their attack on evolution. A small sampling:

> Transitional series *must* have existed in the past, if evolution is true, and the fossil record should reveal at least some of these. . . . The fact

> is, however, that no such transitional series—or even occasional transitional forms—have ever been found in the fossil record.
>
> [Morris 1977, p. 30]

> If continuous evolution is a universal law of nature, as the evolutionist claims, then there should be an abundance of evidence of continuity and transition between all the kinds of organisms involved in the process, both in the present world and in the fossil record. Instead we find great gaps between all the basic kinds, and essentially the same gaps in the fossil record that exist in the modern world.
>
> [Morris 1967, p. 34]

> The history of life reveals a remarkable absence of the many transitional forms demanded by the theory [of evolution]. There is, in fact, a *systematic* deficiency of transitional forms between the higher categories, just as predicted by the creation model.
>
> [Gish 1979, p. 70]

Many more similar quotations could be given. The "absence of transitional forms" argument could well be the paradigm example characterizing the entire creation-evolution controversy for it illustrates how creationists have taken an extremely complex scientific question, simplified the matter to the point of misrepresentation, and then have promoted the blatantly false claim that the fossil record supports the creationist world view. Indeed, from the manner in which creationists have discussed this issue, one can only conclude that either the creationists have consciously adopted the tactic of outright distortion or they are so abysmally ignorant of the scientific arguments and data that their apparent distortions are only accidental, not purposeful. In the short space available, I will attempt to summarize the problem.

After the publication of *On the Origin of Species* in 1859, paleontology assumed an increasingly pivotal role in documenting the evolutionary history of life, and examination of the fossil record itself was perceived as the most direct and efficacious method of historical analysis. At the time that Darwin was formulating his ideas on evolution, nature was being interpreted from a creationist point of view: species were seen as immutable "created kinds" and therefore as discrete entities. In posing an alternative to this world view, Darwin suggested that species are not discrete, that what we observe presently as discrete species are only time-slices of an evolutionary continuum, and that there is a gradual transformation in phenotype through time. Placed in its historical context, this was a logical alternative to a theological interpretation of nature. Most importantly, the idea of a slow, gradual transition in form through time appealed to paleontologists, and ever since the search for these morphological transitions has been an important component of paleontological research (Eldredge and Gould 1972; Cracraft 1979). The contemporary paleontological view about the tempo of evolutionary change, and creationists' misunderstanding of it, will be discussed by Godfrey elsewhere in this book. Here, I want to focus on the nature of intermediate taxa, which are said to provide an evolutionary link between major groups.

Creationists have adopted three lines of argumentation against the existence of transitional forms: (1) they quote liberally from various paleontologists as to the paucity of transitional forms; (2) they define the concept of "transitional form" in a way that is distinctly different from the evolutionists' use of the term; and (3) they simply deny the existence of intermediate taxa, while ignoring a vast scientific literature opposing their position. I will discuss these in order.

In using selected quotations of paleontologists to buttress their own position, creationists have unwittingly entered one of the most controversial theoretical and methodological debates in contemporary paleontological systematics. Inasmuch as this debate has ensued for over a decade in the scientific literature, it is surprising that the creationists have not mentioned its existence (either the creationists are unfamiliar with the scientific literature or they have failed to understand the importance of that literature or they have simply chosen to ignore the problem and adopt a strategy that promotes their theological, not scientific, position). The debate centers on the scientific methods used to postulate and test hypotheses of ancestral-descendant relationship. Traditionally, paleontologists, including most of those quoted by the creationists, have had a conviction that the stratigraphic position of fossil taxa is a primary criterion with which to postulate ancestral-descendant relationships, whereas recent critics of this methodology have stressed the importance of a critical analysis of morphological characteristics (a summary of this literature can be found in Cracraft and Eldredge 1979; Eldredge and Cracraft 1980; and Wiley 1981). If the stratigraphic position of a fossil is an important criterion for recognizing it as an ancestor, it should come as no surprise that it would be extremely difficult to find a specific fossil species that is both intermediate in morphology between two other taxa and is also in the appropriate stratigraphic position. This is no doubt the reason for many of the quotes cited by the creationists about the prevalence of gaps, but other citations are distortions, tailored to suit the creationists' own purposes. For example, in 1972 Schaeffer, Hecht, and Eldredge published an influential paper in which they were critical of paleontological methodology about the construction of ancestral-descendant hypotheses. In support of his argument that there are no transitional forms, Gish (1979, p. 169) quoted from a review of that paper:

> Three paleontologists (no less) conclude that stratigraphic position is totally irrelevant to determination of phylogeny and almost say that no known taxon is derived from any other.
>
> [Van Valen 1973, p. 488]

Although the Van Valen quote gives the appearance of support for Gish's arguments against transitions, a reading of Schaeffer et al. (1972) shows that Van Valen is overstating their position. They clearly do not believe stratigraphy is "totally irrelevant" for examining ancestral-descendant hypotheses nor do they deny the possibility of identifying ancestral species. Rather than engage in a critical analysis of the scientific issues raised by Schaeffer et al., Gish prefers to use Van Valen's statement in a highly biased manner. Gish's unfamiliarity with the scientific literature adds irony to this example: Van Valen, perhaps more than any other contemporary paleontologist, has postulated innumerable phylogenetic connections among fossil taxa and thus offers the poorest support for Gish's viewpoint of anyone he could have misquoted.

Consider another example of distortion. In his discussion of gaps between major groups and the lack of apparent transitions, Gish states:

> It cannot be emphasized too strongly that even evolutionists are arguing among themselves whether these major categories [*sic;* actually they are higher taxa, not categories] appeared *instantaneously* or not! It is precisely the argument of creationists that these forms *did* arise *instantaneously* and that the transitional forms are not recorded because they never existed! [1979, pp. 165–66; italics in original]

Because he is "strongly emphasizing" the debate among evolutionists over the rapidity of appearance of major groups of organisms, Gish should be familiar with some of the literature on the subject; yet he repeatedly misrepresents the concept behind "geologically instantaneous" to make it sound as if it is equivalent to "instantaneous creation." When paleontologists refer to the "instantaneous" origin of a major group, they mean that its appearance in the fossil record can be discerned over a time span that is very short relative to the entire record but that actually may span many millions of years in absolute time; furthermore, paleontologists emphasize that the component taxa of major groups do not all appear at once, as the creationist viewpoint would demand, but only that some of the major defining characters of the group are recognizable in the beginning (logically, this must be true or those fossils would not be assigned to the group in question). Gish knows creationists use "instantaneous" in a biblical sense (Gish and his colleagues at the Institute for Creation Research use six literal days), and he should know that he is misquoting paleontologists.

Creationists also define "transitional form" in such a way as to eliminate the possibility of ever finding one, and it is also a definition that evolutionists would find biologically unacceptable. They appear to envision a transitional form as one that is intermediate in all aspects of form between its immediate ancestor and its descendant. Consider this statement:

> At the very least, there must have been a tremendous number of transitional forms between *Archaeopteryx* and its imaginary reptilian ancestor. Why does not one ever find a fossil animal with half-scales turning into feathers, or half–fore limbs turning into wings?
>
> [Morris 1974, p. 85]

It is possible that creationists do not mean to imply that a transitional form should be intermediate in all respects, but if they do not mean this, then the only reasonable alternative is that transitional forms could be intermediate in *some* respects. If creationists admit this latter possibility, then there would be no reason not to expect the existence in the fossil record of a series of forms, each combining primitive characters inherited from its ancestors, with derived characters that are shared with other members of its lineage. Most modern evolutionists view the question of transitional forms in just this way; nevertheless, it is not difficult to find isolated statements about the lack of intermediate forms. But the use of quotations is not an appropriate way to decide scientific issues, and if one examines the fossil record objectively, there is no doubt that intermediate taxa—mosaics of primitive and derived characters—exist for many major groups. Thus, for example, although *Archaeopteryx lithographica* does not appear to be a direct ancestral species of other birds, it *is* morphologically intermediate in many respects between reptiles and modern birds (Ostrom 1976). A second example: the evidence for a transition between cynodont reptiles and mammals is documented by intermediate fossil taxa, even though this is often denied by creationists (for a summary of the evidence, see Crompton and Jenkins 1979).

In summary, contrary to the protestations of the creationists, intermediate taxa exist for many groups. This does not necessarily mean, of course, that we have highly corroborated hypotheses about direct ancestral-descendant relationships of these taxa. Such a hypothesis is methodologically difficult to study. Nevertheless, when the phylogenetic patterns of the morphologically intermediate taxa are interpreted using modern comparative methods, these fossils present considerable evidence refuting creationist claims and supporting the evolutionary hypothesis.

Biogeography

The distribution of organisms has always played a central role in man's understanding of the history of life. Prior to the eighteenth century not much was known about the different kinds of organisms and their distribution around the world. Most of the knowledge was based on the ecology of those organisms living in Europe and neighboring regions. Because of their biblical background, naturalists of this time generally believed that the species they observed around them would be distributed pretty much worldwide, each according to its own special habitat requirements.

Once expeditions began exploring the world and knowledge accumulated, it did not take long to make the following observations, still true today: (1) there exists an incredible diversity of different kinds of organisms distributed in many areas of the world; (2) each area has its own unique kinds of organisms (it is rare, relatively speaking, for species to be very widely distributed); and (3) species in one area often show similarity relationships to species in a distant area rather than to species in geographically closer areas.

These observations formed the basis for much scientific questioning; they constituted a pattern that biologists realized needed an explanation. The predominant explanation, promoted by many preevolutionary as well as postevolutionary biologists, was that organisms dispersed over the globe, developed differences in their respective areas, and then dispersed to other areas. Dispersalism has been, until quite recently, the primary explanation used by evolutionists (for a general historical summary, see Nelson 1978; and Nelson and Platnick 1981).

Modern biologists have always looked upon biogeography as one of the main sources of evidence for the evolutionary hypothesis. Indeed, biogeography presents such strong support for evolution that creationists have simply ignored the evidence rather than concocting outlandish stories based on revelation (major creationist writings not discussing biogeography include Morris 1974; Wysong 1976; Morris 1977; and Gish 1979). The reasons why biogeographical data present so many problems for the creationists are obvious: (1) they have to explain the great diversity of organisms; (2) they have to explain how all these different kinds of organisms were able to find their way to Noah so that they could avoid the Flood; (3) they have to explain how these different organisms found their way back to their respective areas (from high up on old Mount Ararat, no less); (4) they have to explain why most species (and higher taxa) are endemic to a restricted geographic area when a creationist scenario, in which taxa are dispersing from Ararat, would predict that most species would be widely distributed; and (5) they have to explain why widely separated areas seem to share related kinds of organisms.

From this list of problems, a critical reader should have no difficulty understanding why it is that creationists have ignored the evidence of biogeography. Either they must rely on revelation, pure and simple, or they are compelled to erect a "scientific explanation" that so strains one's credulity it makes them look patently silly. But creationists have an answer for everything. Consider the following examples, from Whitcomb and Morris (1961, p. 87), of their inexhaustible ability to rise to the occasion:

1. Whitcomb and Morris dismiss much of the data of biogeography merely by *ad hoc* argument, claiming that there was a "difference of

climatic and zoogeographical conditions before the Flood as compared to the postdiluvian area" (1961, p. 87).

2. And how did the animals get to the Ark and survive for more than three months (away from their native habitats)? By "the possible impartation of migratory instincts and powers of hibernation to the animals by God with respect to the gathering and caring for the animals during that year of cosmic crisis" (1961, p. 87).

3. What about after the Flood? Simple: "It is by no means unreasonable to assume that all land animals in the world today have descended from those which were in the Ark" (1961, p. 87).

4. And how did these animals distribute themselves from Mount Ararat? Whitcomb and Morris, as do virtually all good creationists, offer a simple answer: "It would not have required centuries even for animals like the edentates to migrate from Asia to South America over the Bering land bridge. Population pressures, search for new homes, and especially the impelling force of God's command to the animals kingdom (Gen. 8:17) soon filled every part of the habitable earth with birds, beasts, and creeping things" (1961, p. 87).

Whitcomb and Morris wrote their book at a time when even many professional biogeographers clung to the idea that distribution patterns could be explained by massive waves of dispersal. Nevertheless, the most ardent dispersalist would never accept the idea that all the animals dispersed from Ararat to the far corners of the globe, and differentiated into a myriad of forms, within a few thousand years. That scenario is childish myth—it is fundamentalistic religion, not science.

During the last decade biogeographers have come to realize that when the postulated phylogenetic relationships of organisms—both plants and animals—are examined relative to their distributions, many highly congruent, nonrandom patterns emerge. Many of the taxa endemic to tropical South America have as their closest relatives taxa endemic in Africa; taxa endemic to the cool temperate regions of southern South America have as their closest relatives taxa endemic to New Zealand and the cool temperate regions of Australia and Tasmania; and some taxa endemic to western North America have as their closest relatives taxa endemic to China and eastern Asia; and so on. Given these many patterns of distribution, it has become apparent that biotas now separated were once connected and that after separation the individual components (taxa) of these biotas differentiated in isolation. In other words, dispersal from one area to another, followed by differentiation, is apparently not as important as once thought. And as biogeographers study the problem in more detail, it is becoming apparent that these patterns of biotic separation are correlated with changes in earth history, continental drift being the most obvious example. The literature on this new approach to distribution, called vicariance biogeography, is already very large and promises to change many of our ideas about the history of life (see Platnick and Nelson [1978]; Nelson and Platnick [1980, 1981]; Cracraft [1980]; and Nelson and Rosen [1981]).

While professional biologists struggle with the enormous complexity of the systematic relationships of organisms, their distribution patterns, and the correlation of those patterns with hypotheses about earth history, creationists take the easy road to "knowledge" and simply force a biblical interpretation on nature. The most detailed biogeographic analysis by a creationist in recent years is that of Howe (1979), who attempts to "explain" the distribution of angiosperms. Let's examine his method of analysis:

1. Howe informs us (p. 38) that plants present "an array of unrelated types," a ludricrous assertion in this day and age. If Howe were working in a monastery in the fourteenth or fifteenth century he might be excused, but no modern scientist with any competence could make that statement. Of course, he ignores the large literature on plant systematics.

2. He proclaims (p. 41) that in angiosperms the genus is probably the "created kind," but presents no evidence informing us how he made that decision.

3. He interprets (pp. 40–43) endemics as having been isolated by Flood-related events, particularly by continental drift, which is said to have occurred *after* the Flood. This is so manifestly silly as to require little comment. Any person claiming that the vast literature on plate tectonics and continental drift supports a time scale for these events on the order of a few thousand years is suffering a delusion of religious faith and is definitely not gifted with any faculties for scientific reasoning.

Howe is a typical example of extreme creationist thinking and reasoning when it comes to biogeographic data: ignore the evidence, claim that the evidence (which you have ignored anyway) fits a literal interpretation of Genesis, and then claim that what you are doing is science. Who can take such a view seriously?

Classification

In biology, formal classifications have existed primarily as a means of expressing the content of natural groups. Since Darwin, most biologists have taken these groups to be those descended from a common ancestor. The precise content of natural groups is a scientific hypothesis. Thus, alternative hypotheses could be proposed for the content of natural groups. It might be supposed, to mention an example used earlier, that whales should be classified with sharks rather than with mammals. Also, a hypothesis about the content of a natural group is predictive and therefore testable. These hypotheses make predictions about the hierarchical distribution of similarities shared among the organisms in question. Classificatory hypotheses such as this have been made for a very long time and have been discussed in detail in the systematic literature (see especially Eldredge and Cracraft 1980; Nelson and Platnick 1981; and Wiley 1981).

Because systematists are scientists, there are bound to be considerable differences of opinion about the theory and methodology of biological classification. A large literature exists on the subject, and differences of opinion continue to be expressed in the technical journals, including *Systematic Zoology* and *Systematic Botany.* Creationists have commented on classification, but it is obvious from their statements that they are not interested in the scientific aspects of classification, but rather in how that literature can be used to support creationism. Some typical comments:

> Actually, the more confusion in classification—the less organisms are distinctly demarcated—the more it will appear as though organisms blend into one another, i.e., the more evolution will appear true.
>
> [Wysong 1976, p. 58]

> The fact that categories of natural phenomena can be arranged in orderly classification systems (table of chemical elements, biological taxonomy as in the Linnaean system, a hierarchy of star types, etc.) is a testimony to creation. That is, if all entities were truly in a state

> of evolutionary flux, classification would be impossible. In biological classification, for example, it would be impossible to demark where 'cats' leave off and 'dogs' begin. [Morris 1974, pp. 21–25]

The arguments of Wysong and of Morris and his colleagues are straw men of their own manufacture. Only the creationists claim that there must be an unbroken, gapless chain of morphology if evolution is true —evolutionists certainly do not think this; consequently, their arguments vanish. Inasmuch as they provide little indication of having read the scientific literature on classification theory, they are probably unaware that many biologists, who nevertheless accept evolution, advocate classifications that do not necessarily express evolutionary relationships. There are thus many ways to classify.

Most biologists, however, do advocate that classifications be expressive of phylogenetic relationships to a greater or lesser degree, and the creationists use these classifications themselves. If they believe these are arbitrary (e.g., Morris 1974, pp. 71–72), one would think they would prefer their own classification to that of the evolutionists. And if they attempted such a classification, chances are they would adopt most of the same groups as the evolutionists. Why should that be? Because these groups share defining similarities: the derived characters of the evolutionist. Linnaean classification schemes, used throughout biology, express the pattern of hierarchical similarity discussed earlier, and the reason seems to be that nature is hierarchical—a prediction of evolution, not creationism.

Discussion

It is a depressing situation when biologists of the late twentieth century find it necessary to answer challenges to evolutionary science by a small group of religious zealots. This is especially so since the issues are not really scientific in content, but rather are sociological, political, and economic. The arguments of the modern-day creationists hark back to a time, hundreds of years ago, when superstition and religious dogma controlled—or attempted to control—peoples' lives and intellectual activities: it is easier to believe the truth of religious dogma than to seek rational understanding, easier to believe than to question, easier to be told what to know than to decide for oneself.

I say that the present controversies are sociological and political rather than scientific because the creationists evidence so little understanding of modern biology, and evolutionary biology in particular, that they could not be expected to carry on an intelligent dialogue on the scientific issues. The main reason for this, of course, is their commitment to conservative, religious fundamentalism, rather than to the pursuit of knowledge. Naturally they will seek a degree of understanding only sufficiently deep to accept that which reinforces their own psychological and philosophical belief system. In the final analysis, all arguments of the creationists come back to religious dogma, and despite an intense political campaign to convince the public that they are unbiased, critical scientists, it is common knowledge, at least among scientists, that that campaign is founded almost entirely on their attempts at establishing the legitimacy of their self-manufactured rubric "scientific creationism." This judgment is not intended to question the sincerity of the motives or beliefs of the many nonscientists who undoubtedly evaluate the world around them from deep, personal supernaturalistic faith. The leadership of the "scientific creationist" movement, on the other hand, frequently has exceeded the boundaries of this faith to proselytize creationism by criticizing evolutionary biology with innuendo, distortion, and outright deception. Such has to be the judgement of their writings on evolution, unless one wants to accept the only viable alternative hypothesis: that these writings manifest a sincere form of religious zeal and ignorance of the scientific issues.

In this chapter I have tried to show how creationists use the data of systematics and comparative biology. One might ask how this use can be characterized from a scientific standpoint, and the answer would be unequivocal: superficial and unprofessional. This may or may not be unexpected, depending upon how much time one spends critically reading their writings rather than listening to their self-proclamations about their credentials and qualifications. Creationists complain incessantly about the impossibility of getting their writings published in the scientific literature (many people do not realize that virtually all their publications are produced by their own organizations). The truth of the matter is that, if the writings quoted here are characteristic of their understanding of the scientific issues, the creationists would seem to have nothing to say that is of any scientific value. It is their lack of competence, rather than prejudice on the part of evolutionists, that prevents them from publishing. Indeed, there are many devout Christians who regu publish in the scientific literature, but they promote competent science, not religion disguised as science.

As I have demonstrated in this chapter, creationists offer explanations of natural phenomena (distribution patterns, for example), the only content of which is strict adherence to and belief in the literal truth of Genesis. This immediately disqualifies these explanations as being scientific. The use of the term "scientific creationism" is a charade, and is merely the political propaganda of a conservative, fundamentalist belief system.

Creationism could easily be ignored as a manifestation of a religious cult or fringe group if it were not for its potential to become more broad-based within Christian fundamentalism. Those creationists cited in this chapter are extremists; they cannot be allowed to convince sincere Christians that their religious faith will be destroyed by knowledge derived from science and the humanities. Personal choice and belief will maintain that faith, if one so chooses. But the attitudes of the creationists constitute a threat to critical inquiry and to the growth of knowledge. The reason for this is simple: as long as you think you have the truth, you will feel no need to pursue knowledge and understanding. The creationists have their truth, and they want to convey that truth through indoctrination:

> Only the Creator—God himself—can tell us what is the truth about the origin of all things. And this He has done, in the Bible, if we are willing simply to believe what He has told us. [Morris 1967, p. 20]

> If we expect to learn anything more than this about the Creation, then God above can tell us. And He has told us! In the Bible, which is the Word of God, He has told us everything we *need* to know about the Creation and earth's primeval history.
> [Morris 1967, p. 54; italics in original]

> The final and conclusive evidence against evolution is the fact that the Bible denies it. The Bible is the Word of God, absolutely inerrant and verbally inspired. . . . The Bible gives us the revelation we need, and it will be found that all the known facts of science or history can be very satisfactorily understood within this Biblical framework.
> [Morris 1967, p. 55]

These are the words of Henry Morris, director of the Institute for Creation Research, perhaps the most influential creationist organization. If one agrees with him, then what I have written will not make much sense or difference. If, on the other hand, one does not necessarily want Henry Morris to decide what one needs to know, then perhaps what I have written will be useful in evaluating the cogency of creationist thinking.

REFERENCES CITED

Alberch, Pere. 1980. Ontogenesis and morphological diversification. *American Zoologist* 20:653–67.

Cracraft, Joel. 1979. Phylogenetic analysis, evolutionary models and paleontology. In *Phylogenetic analysis and paleontology,* ed. Joel Cracraft and Niles Eldredge, pp. 7–39. New York: Columbia Univ. Press.

—. 1980. Biogeographic patterns of terrestrial vertebrates in the southwest Pacific. *Palaeogeography, Palaeoclimatology, Palaeoecology* 31:353–69.

—. 1981. The use of functional and adaptive criteria in phylogenetic systematics. *American Zoologist* 21:21–36.

Cracraft, Joel and Eldredge, Niles, eds. 1979. *Phylogenetic analysis and paleontology.* New York: Columbia Univ. Press.

Crompton, A. W. and Jenkins, Farish A., Jr. 1979. Origin of mammals. In *Mesozoic mammals: the first two-thirds of mammalian history,* eds. Jason A. Lillegraven, Zofia Kielan-Jaworowska, and William A. Clements, pp. 59–73. Berkeley and Los Angeles: Univ. of California Press.

Darwin, Charles. 1859. *On the origin of species by means of natural selection, or the preservation of favored races in the struggle for life.* Facsimile edition (1975). Cambridge, Mass.: Harvard Univ. Press.

Eldredge, Niles and Cracraft, Joel. 1980. *Phylogenetic patterns and the evolutionary process.* New York: Columbia Univ. Press.

Eldredge, Niles and Gould, Stephen J. 1972. Punctuated equilibria: an alternative to phyletic gradualism. In *Models in paleobiology,* ed. T. J. M. Schopf, pp. 82–115. San Francisco: Freeman, Cooper and Co.

Gish, Duane T. 1979. *Evolution? The fossils say no!* 3rd ed. San Diego: Creation-Life Pubs.

Gould, Stephen J. and Lewontin, Richard C. 1979. The spandrels of San Marco and the Panglossian paradigm: a critique of the adaptationist programme. *Proceedings of the Royal Society of London* 205B:547–65.

Hennig, Willi. 1966. *Phylogenetic systematics.* Urbana, Ill.: Univ. of Illinois Press.

Howe, G. F. 1979. Biogeography from a creationist perspective. 1: taxonomy,

geography, and plate tectonics in relation to created kinds of angiosperms. *Creation Research Society Quarterly* 16:38–43.
Lack, David. 1947. *Darwin's finches.* Cambridge, Eng.: Cambridge Univ. Press.
Lammerts, William E. 1970. The Galapagos Island finches. In *Why not creation?*, ed. William E. Lammerts, pp. 354–66. Grand Rapids, Mich.: Baker Book House.
Løvtrup, Søren. 1974. *Epigenetics.* New York: Wiley.
Marsh, Frank Lewis. 1947. *Evolution, creation and science.* Washington, D. C.: Review and Herald Pub. (not seen; quoted by Wysong 1976, p. 59).
Mayr, Ernst. 1963. *Animal species and evolution.* Cambridge, Mass.: Harvard Univ. Press.
Morris, Henry M. 1967. *Evolution and the modern Christian.* Philadelphia: Presbyterian and Reformed Pub. Co.
—. 1974. Ed. *Scientific creationism.* San Diego: Creation-Life Pubs.
—. 1977. *The scientific case for creation.* San Diego: Creation-Life Pubs.
Nelson, Gareth J. 1978. From Candolle to Croizat: comments on the history of biogeography. *Journal of the History of Biology* 11:269–305.
Nelson, Gareth J. and Platnick, Norman I. 1980. A vicariance approach to historical biogeography. *Bioscience* 30:339–43.
—. 1981. *Systematics and biogeography: cladistics and vicariance.* New York: Columbia Univ. Press.
Nelson, Gareth J. and Rosen, Donn E., eds. 1981. *Vicariance biogeography: a critique.* New York: Columbia Univ. Press.
Ostrom, J. H. 1976. *Archaeopteryx* and the origin of birds. *Biological Journal of the Linnean Society* 8:91–182.
Parker, Gary. 1980. *Creation: the facts of life.* San Diego: Creation-Life Pubs.
Platnick, Norman I. and Nelson, Gareth J. 1978. A method of analysis for historical biogeography. *Systematic Zoology* 27:1–16.
Schaeffer, B., Hecht, M. K., and Eldredge, N. 1972. Phylogeny and paleontology. *Evolutionary Biology* 6:31–46.
Siegler, Hilbert R. 1978. A creationists' taxonomy. *Creation Research Society Quarterly,* 15:36–38.
Van Valen, Leigh. 1973. Review of *Evolutionary Biology,* vol. 6. *Science* 180: 488.
Whitcomb, John C., Jr. and Morris, Henry M. 1961. *The Genesis flood.* Grand Rapids, Mich.: Baker Book House.
Wiley, Edward O. 1981. *Phylogenetics: the theory and practice of phylogenetic systematics.* New York: Wiley.
Wysong, R. L. 1976. *The creation-evolution controversy.* Midland, Mich.: Inquiry Press.

11

Creationism and Gaps in the Fossil Record

BY LAURIE R. GODFREY

Taphonomy and the Assessment of Missing Data

I vividly recall my first exposure, in my student days, to the science of taphonomy. I was in the osteology laboratory on the fifth floor of Harvard's Peabody Museum of Anthropology. A paleontology student from the neighboring Museum of Comparative Zoology entered the room, strode over to our freezer, and removed half of a freshly frozen monkey head. I immediately became exceedingly curious and feigned possessiveness. "What are you doing with *my* monkey?" I inquired. "Come with me and you'll see," he answered.

I followed him to a laboratory in the Museum of Comparative Zoology. There, behind the locked door, was a young crocodile. My friend tossed the monkey head to the crocodile, who disposed of it forthwith. I was initially dismayed by what seemed a senseless waste of anthropological resource materials, but after being escorted to the fossil preparation room in the basement of the Museum of Comparative Zoology, I understood the purpose of my friend's Peabody escapade. A large block of sedimentary rock that had been removed from a fossil crocodile bed sat in the middle of the room. The rock was filled with minute pieces of teeth and bone of ancient crocodiles and their victims. In order to reconstruct the community that once lived there—the "biocoenose" or "living community"—my friend had to assess what was missing from the "thanatocoenose" or "death assemblage" that had been preserved. The crocodile was fed a diet of whole animal parts (with bones and teeth intact) because the student required some data on this animal's natural digestive processes. One of his tasks would be to measure the relationship between those tissues ingested and those excreted by a crocodile. With these and other experimental data, and with careful inch-by-inch analysis of the fragmentary remains in his sample of ancient crocodile bed, he hoped to reconstruct long gone populations of both predators and prey.

That was my introduction to a field that since has grown into a significant subdiscipline of geology (see, for example, Behrensmeyer and Hill 1980; Shipman 1981). Its name derives from the Greek "taphos," which means "tomb" or "burial," and "nomos," which means "laws." The goals of the taphonomist are formidable; many diverse factors affect the "life history" of organisms after death. Cadavers may be attacked by armies of ants or vertebrate scavangers; they may be weathered, trampled, or transported and sorted by moving water. Those bones that are buried before they are completely destroyed will be altered by the processes of fossilization and may be badly deformed by postdepositional pressure. The taphonomist approaches the sedimentary record of past life with slow and deliberate care, seeking to extract the maximum possible information from every scrap of evidence—from the smallest hand bone or tooth fragment to the most perfect skull.

It is a far cry from the paleontology of yesteryear—the heyday of the great bone collectors—when men like Edward Drinker Cope and Othniel Charles Marsh gathered their equipment and teams of hired help and raced across the vast American West searching for great "finds," each zealous in his efforts to outwit the other.

Paleontologists of the late 1800s had virgin fossiliferous territories to explore. Central North America contained riches from the Cretaceous epoch (more than 65 million years ago), when the central portion of the continent was covered by a shallow sea extending from the Arctic to the Gulf of Mexico. Eager bone hunters could recover the remains of spectacular plesiosaurs and pterosaurs that swam in and flew over the Cretaceous sea, as well as of the dinosaurs that roamed its ancient shores. Furthermore, the Bridger basin of Wyoming and Utah yielded the remains of the early Tertiary mammals who inherited the land after the dinosaurs had become extinct. The stories these fossil hunters left behind are legion (see Lanham 1973)—stories of jealousy, errors of judgment (Marsh put the wrong head on brontosaurus, for example, an error that confounded paleontologists until very recently), adventure, even murder (Lanham 1973, p. 111). Periodically, conflicts arose over taxonomic names. All hell broke loose when Cope and Marsh respectively assigned the names *Tomitherium rostratum* and *Limnotherium affine* to the same Eocene primate, because there was no unambiguous way to choose between them. Ordinarily, the "law of taxonomic priority" prevents such ambiguity by requiring the selection of the name first published in the scientific "literature," provided that name is not already the valid name of another organism. But in this case there was no first-published name; both were published in journals bearing the date 7 August 1872. So Cope declared his name preferable on the grounds that his description was longer by several lines and his measurements greater by one. Fortunately, it was later discovered that senior paleontologist Joseph Leidy had already named the animal in dispute. Leidy's *Notharctus* is the name we remember today.

How strange my friend's crocodile quarry project would have seemed to explorers like Cope and Marsh! Indeed, the notion of taphonomy was not conceived until 1940 when Efremov stated his concern for understanding the "transition (in all its details) of animal remains from the biosphere to the lithosphere" (1940, p. 85). But taphonomists are explorers too, and I would judge their discoveries as exciting as any made in the 1800s. Taphonomists are, in a sense, discoverers of the earth's *missing* data; they can assess contexts invisible to all but their highly trained eyes. They can recognize the telltale signs of postmortem gnawing, of trampling, of slow or rapid water transport, of oxidizing or reducing depositional environments, of physical and chemical weathering, of postdepositional deformation, and so on. They can tell you why no shoulder blades, vertebrae, hand bones, and foot bones may be represented in a deposit loaded with skulls, jaws, and occasional long bones of fossil vertebrates.

Functional anatomists fill in other aspects of missing data. Soft tissue is only preserved under unusual circumstances, but functional anatomists working with skull endocasts can reconstruct the surface anatomy of the brain. Working with rugosities and depressions on bones they can reconstruct some muscle attachments. Working with joint articular surfaces, they can reconstruct the "movement potential" of the bones (see, for example, Hildebrand 1974; Morbeck, Preuschoft, Gomberg 1979).

Paleontologists today are also explicitly concerned with processes of evolutionary change and the implications of paleontological data for evolutionary theory (Eldredge and Gould 1972; Gould and Eldredge 1977; Hallam 1977; Stanley 1979; Gould 1980). This is another example of how paleontologists draw inferences from both what *is* and what *is not* represented in the fossil record—the spatial and temporal patterns of continuity and discontinuity in form.

We can infer much from the fossil record that we can only partially see—lifeways and environments of extinct organisms, temporal changes in the diversity of taxa, processes of evolutionary change. We can even guess how much information is missing completely from the fossil record. We know, for instance, that almost all organisms lacking hard parts live and die without leaving a trace and that the same is true of the great majority of organisms *with* hard parts. We know that many organisms that are buried and preserved disappear because their fossilized remains are eroded. For relatively easily preserved taxa such as sea urchins, paleontologists may have recovered remnants of half of the species that ever lived (Stanley 1981), despite these remnants' great geological age. For other long-lived groups, our recovery rate is far lower. In general, our record of the last 5 or 10 million years of earth's history is superior to that of earlier times. This is one of the reasons why the human fossil record is comparatively excellent. Ancient deposits not destroyed by denudation may be destroyed by metamorphism. Yet erosion alone wipes out much of our record of past life on earth. Whenever conditions exist that favor deposition and burial in one place, there will be conditions favoring erosion somewhere else. Like burial processes, the earth's erosional processes filter out all but a minor portion of the total "potential" data. These processes both reveal and destroy the fossilized remains of prehistoric organisms. Fossils that have been sheltered in a rock matrix for many millions of years become exposed by erosion and are assaulted by natural weathering processes once again. They may "live" at the surface, ready to tell their exciting stories of life and death, for but an instant of geologic time.

Charles Darwin bemoaned the data missing from the fossil record. He borrowed a metaphor from geologist Charles Lyell and described the scarred surface of the earth as the lone remaining volume of a several volume set. And in that volume, only a few chapters remain. In those chapters, only a few lines remain, and these are written in a language that has been modified with the slow passage of time.

A century and a quarter after Darwin, the fossil record is still incomplete. It will *always* be incomplete. But we know far more about the life of the past today than we did in Darwin's day. The fossil record has provided students of life with a valuable diachronic glimpse of life on earth and with independent corroboration of life's evolution.

The Fossil Record and the Evolution of Life on Earth

Darwin drew upon many independent lines of evidence in formulating and defending his thesis of descent with modification by natural selection: artificial selection, natural variation and hybridization, fossils, geographic distribution of living species, comparative morphology, and comparative embryology. The fossil record was not primary among these; indeed, Darwin spent as many pages excusing problems raised by the fossil record as he did drawing from that record for factual support. "What a paltry display we behold!" he declared of the fossil collections in the best of paleontological museums (1859, p. 294). "I do not pretend that I should ever have suspected how poor was the record in the best preserved geological sections, had not the absence of innumerable transitional links between the species which lived at the commencement and close of each formation, pressed so hardly on my theory" (p. 307). After reviewing several examples of the "sudden appearance" of whole groups of organisms, he concluded, "Those who believe that the geological record is in any degree perfect, will undoubtedly at once reject the theory [mutability of species by natural selection]" (p. 315).

Of course, the fossil record known to Darwin did contain clear evidence of evolution. The order of appearance of fossil forms in geological strata is far from random with respect to morphology. Darwin recognized the successive appearance and diversification of allied groups of organisms as evidence for evolution. But he clearly expected *more* of the fossil record than this. Natural selection, in his view, was an agent of slow and steady change within lineages. The fossils should therefore document "innumerable" gradational links between ancestral and descendant forms. Darwin explained the rarity of transitional links in two ways. First and foremost, erosional hiatuses and periods of nondeposition interrupt gradational series: "If such gradations were not all fully preserved, transitional varieties would merely appear as so many new, though clearly allied species" (1859, p. 306). Second, taxonomic practice stymies our recognition of series since taxonomists require that each organism be assigned some species name, and continuous series will therefore necessarily be artificially divided: "These links, let them be ever so close, if found in different stages of the same formation, would, by many paleontologists, be ranked as distinct species" (1859, p. 307).

Darwin did not pose biological explanations for the sudden appearances of species or groups of species in the fossil record. His understanding of genetic change precluded such analysis. He never doubted the existence of continuous series of transitional forms. Nor did he consider it likely that transitions could be temporally ephemeral. Twentieth-century paleontologist George Gaylord Simpson (1944, 1953) departed from Darwin's pervasive gradualism when he suggested that many major transitions might involve abnormally rapid rates of biological evolution. Simpson called this "quantum evolution." But to Darwin (1879), if flowering plants (angiosperms) appear "suddenly" in the Cretaceous epoch, it must be because bad luck has concealed their long and gradual early history. It could not be because the origin of angiosperms was itself biologically revolutionary in nature. Darwin preferred lost continents to biological revolutions.

Of course, paleontologists today have far more data upon which to base their analyses of transitions—data bearing directly on many of Darwin's fossil enigmas. Resolutions of these puzzles have not always matched Darwin's expectations, however. We have found no lost continents; radiometric dating has revealed no major lost segments of geologic time. We *have* found clear evidence of transitional links between major vertebrate groups (see Hinchliffe and Johnson 1980; Panchen 1980; Schultze 1977; and Szarski 1977 on the origin of tetrapods; Ostrom 1975, 1976, 1979; and Feduccia 1980 on the origin of birds; Lillegraven, Kielan-Jaworowska, Clemens 1979; Jenkins and Parrington 1976; and Crompton and Parker 1978 on the origin and early evolutionary history of mammals). We *have* found sources in Pre-Cambrian rocks for the famous Cambrian explosion of multicellular life forms (see Cloud

1977; Valentine 1977; and Valentine, in press). We *have* found primitive antecedents of the flowering plants—links between the angiosperms and the gymnosperms that dominated the land before angiosperms appeared sometime in the Cretaceous (Hughes 1976; Beck 1976). We *have* found links between modern humans and fossil apes (Brace, this volume; Johanson and Edey 1981). In many cases, the problem is not a lack of intermediates but the existence of so many closely related intermediate forms that it is notoriously difficult to decipher true ancestral-descendant relationships. In a very real sense, the fossil record is far better testimony to evolutionary change than Darwin, in his later years, probably imagined possible.

To be sure, there are still major groups whose origins remain enigmatic. Bats, for example, have the poorest fossil record of all major vertebrate groups despite their numerical abundance in the world today. This is not only because bats are small and lightly built; they are also largely forest and cave dwellers. Forest faunas, in general, are poorly preserved, and cave sites older than several million years are very rare. There are some remarkably well preserved early Tertiary fossil bats, such as *Icaronycteris index*, but *Icaronycteris* tells us nothing about the evolution of flight in bats because it was a perfectly good flying bat. But if the origin of flight in bats from some ancestral arboreal insectivore (presumably via a parachuting or gliding stage) remains elusive, the inference drawn from neontological data that bats evolved from some primitive eutherian (or placental) mammal is well supported by fossil bat teeth. Bats are essentially flying insectivores with a dentition modified from the basal eutherian plan. Paleontologist Bob Slaughter (1970) was able to describe in detail evolutionary trends within major bat groups (Megachiroptera and Microchiroptera). More exciting, from our point of view, are his conclusions regarding the probable evolutionary pathways of the earliest bats. Indeed, we now have fossil teeth that document the origins and early dental evolution of all placental mammals. As Slaughter says, "Not only are these [fossil teeth] intermediate in form between molars of extinct [mammalian] orders and primitive members of living orders, but the new material is chronologically intermediate as well", (1970, p. 52).

Slaughter goes on to underscore the significance of these finds when coupled with later evidence of divergent evolutionary trends in different (and still extant) mammalian orders. "One of the most fascinating facts in paleontology is that tooth patterns of shrews, rats, cats, bears, horses, camels, man and *bats* can be traced back through time to a single type of dentition" (1970, p. 52, italics in original).

Early members of living mammalian orders are primitive in cranial and postcranial skeletal features as well as in their dentitions. The order Primates appears "suddenly" in the Cretaceous and Paleocene, but these first primates looked nothing like modern monkeys, apes, or humans. Like bats, early primates were primitive eutherian mammals lacking whole sets of traits that evolved later in subsequent primate adaptive radiations. Details of the morphology of the base of the skull and details of molar cusp morphology tell us that these fossil animals were indeed primates, but they looked much more like arboreal shrews or rodents than like modern monkeys. Their eyes did not point forward; they did not possess nails on all digits instead of claws; they had keen senses of smell, and so on. Despite the poorness of our record of primate origins and the difficulty of reconstructing their precise ancestry, primate fossils very clearly support the inferences that primates underwent a temporal succession of adaptive radiations and that they were ultimately derived (along with other extant mammalian taxa) from a primitive eutherian mammalian stock (see Szalay and Delson 1979; Luckett and Szalay 1975).

Primitive primates shared their world with archaic "carnivores" and primitive "ungulates." But there were no baboons, macaques, chimps, humans, cows, pigs, cats, and dogs living in the Paleocene, the very beginning of the Age of Mammals. Indeed, Simpson (1953) argued that early carnivores and ungulates, taken together, were less varied than some recent carnivore families such as the Mustelidae (e.g., skunks, weasels, badgers, wolverines and others) and the Viverridae (e.g., civets, genets, and the mongoose). The "sudden appearances" of various extant mammalian orders in no way disconfirms their evolution.

Let us briefly examine the early evolution of the flowering plants, problems about which paleobotanists had been confounded by for about a century. Their sudden appearance in the Cretaceous epoch bewildered Darwin, who called it an "abominable mystery" in a letter to Hooker in 1879. As of twenty years ago, angiosperm origins were still poorly understood, and Tom Harris echoed Darwin's sentiments in a speech delivered to the British Association for the Advancement of Science when he asked his colleagues "to look back, not on a proud record of the success of famous men, but on an unbroken record of failure" (1960, p. 207). But in 1976 the picture looked very different. Paleobotanists James A. Doyle and Leo J. Hickey had discovered a sequence of fossil pollens and fossil leaves from the Cretaceous Potomac group of the Atlantic coastal plain. Both leaves and pollens in the lowest beds of the geologic column were very primitive, and younger beds documented the evolution of leaf shapes, margins, and venation patterns characteristic of modern angiosperms, as well as the increasing complexity of reproductive parts. Doyle and Hickey interpreted these plants as members of a primary adaptive radiation of angiosperms, commenting that "the fossil record is now of major significance as evidence for the solution of Darwin's 'abominable mystery' of their origin" (1976, p. 198). While very primitive angiosperm fossil pollens exist elsewhere (in deposits of the next earlier stage), Doyle and Hickey felt that there was "no reason to postulate any extensive prior diversification" (1976, p. 198).

It now seems evident that paleobotanists were long unsuccessful in their search for primitive angiosperms because they were looking in the wrong deposits! Never really believing that angiosperms could originate and diversify in only tens of millions of years, most paleobotanists had focused on deposits from the early Mesozoic or, indeed, late Paleozoic (see Beck 1976 for an overview of the history of paleobotanical research). But the evolutionary origin of angiosperms was apparently much closer to the time of their initial widespread appearance than had been previously thought. In retrospect, we might add that angiosperm paleobiology is still remarkable, though not for its lack of intermediate forms. It is remarkable for its clear documentation of the early adaptive radiation of a major taxonomic group—a biological revolution. As Stanley (1979, 1981) and others (e.g., Eldredge and Gould 1972; Gould and Eldredge 1977) have repeatedly emphasized, if such a mode of origin (involving rapid multiplication coupled with morphological innovation) is typical of the origin of major taxa, there can be little wonder that the fossil record documents a succession of abrupt appearances of taxa (which are themselves elements of larger transitions), rather than innumerable smooth gradations.

Transitions exist at two levels in the fossil record. First, there are species-level transitions. These are the transitions the young Darwin expected to see in the fossil record. They are rare, but they are known. Second, there are intermediates between groups at higher levels of the taxonomic hierarchy: families, orders, classes, phyla. These exist in abundance in the fossil record—fishes with limblike fins and lungs, mammal-like therapsid reptiles, birdlike theropod dinosaurs, hominids combining primitive apelike and derived humanlike traits, and so on. Even when intermediates exist in abundance, some modern paleontologists are loathe to arrange them into ancestral-descendant sequences. Evolutionary trends often occur not in single gradually changing lineages as Darwin envisioned, but as sequences of innovations in complex series of lineages. Trends may be erratic and only roughly directional. To be sure, apparently smooth and gradual changes within lineages have also been documented (Gingerich 1976); these often involve shifts in size and associated changes in shape. Long-term, "macroevolutionary" trends, then, may be products of directional selection acting within lineages (Grant's 1963 "phyletic trends") or of the differential survival of species (Grant's "speciational trends"). The multiplication of lineages through splitting ("speciation"), when coupled with the high improbability of preservation of members of *single* lineages, contributes to the reluctance of many paleontologists to assign ancestral (rather than close cousin) status to particular fossil forms. Only under exceptional conditions of preservation will direct ancestors of successive adaptive radiations be preserved.

"Unbridgeable" Boundaries

Duane Gish (1978) prepared the major creationist account of the fossil record for public school use. The book is provocatively called *Evolution? The Fossils Say NO!* Its treatment of angiosperm origins is typical of its treatment of other paleontological data—misleading, incomplete, full of half-truths and outright falsehoods. Gish disposes of angiosperm evolution in a single paragraph by citing a remark made by botanist E. J. H. Corner in 1961 that despairs of the (then valid) lack of known primitive angiosperms. Gish fails to mention the major books on angiosperm paleobiology that should have been available to him in 1978: Beck's *Origin and Early Evolution of Angiosperms* (1976) or Hughes's *Palaeobiology of Angiosperm Origins* (1976). Anyone serious about surveying the fossil evidence of angiosperm evolution could hardly have missed them.

It seems remarkable that even today, with all that is known of the fossil record, special creationists steadfastly proclaim that paleontological discoveries have vindicated their view that the earth's life forms were created "suddenly" (in no more than six twenty-four-hour periods) and that the subsequent history of living creatures is one of variation and diversification of "created kinds" plus extinction. Why, if all "kinds" were created "simultaneously," do organisms typically appear as allied groups in fixed geological strata? Somehow the very people who ridicule evolutionists for sometimes invoking low probability events in their explanatory schema see no problem in accepting the absurdly unlikely proposition that the morphologically nonrandom succession of life forms in the fossil record is the product of hydrodynamic sorting by a catastrophic flood! Do these people seriously believe that hydrodynamic sorting better accounts for the successive appearance of groups of organisms than do speciation and adaptive radiation? Can they seriously believe that hydrodynamic sorting also accounts for the numerous macroevolutionary trends so well documented in the fossil record? Is it really possible that horses, humans, cows, and rats were true contemporaries of the primitive mammals known from Mesozoic deposits, but somehow only various noneutherian, apparently transitional, and primitive eutherian mammals managed to die in the right places? What hydrodynamic properties of their bones could possibly link these small mammals with the giants of the reptile world?

Despite the evident paradox, creationists love to write about the fossil record. After all, the fossil record has its gaps, and gaps have long been a favorite subject of antievolutionists. Creationist Henry Morris seems to begin with Darwin's dilemma—"There ought . . . to be a continuous intergrading series" (1975, p. 20)—but quickly converts this argument into an argument Darwin never made (and that is patently false)—that the fossil world has revealed no intermediates between "basic kinds" such as "sharks and whales" (1975, p. 20). What nonsense! There are multitudes of intermediates between such "kinds" as Morris cites, both in the paleontological and contemporary worlds. Between sharks and whales, for example, we find bony fishes, amphibians, reptiles, mammal-like reptiles, and some mammals. As Cracraft (this volume) has explained, each intermediate taxon exhibits a set of derived features that distinguish it and its descendants from all other groups of organisms. Amphibians, reptiles, mammallike reptiles, successive groups of primitive mammals, and primitive whales in turn display increasingly derived sets of traits that form the *successive links* binding fishes and modern whales. Morris confuses his audience by constructing a false test of evolution (as did creationist Luther Sunderland when he made the widely publicized humorous slide of a modern cow "evolving" into a modern whale). Evolution does not demand that the "series" connecting any two extant forms will follow the shortest morphological line between them; it demands that any two extant organisms share a common ancestor—an organism that may be very far indeed from the morphological "mean" of the extant pair. In the case of the shark and whale, that common ancestor was a primitive fish! Given numerous speciation events subsequently separating the pair, most members of intermediate taxa will not be in the direct line of ancestry of *either* modern form. They will be "cousins."

The fossil record also documents "stasis"—periods of little or no apparent morphological change, sometimes lasting thousands of years, sometimes millions of years, or sometimes even hundreds of millions of years to produce "living fossils." And now paleontologists Niles Eldredge, Stephen Jay Gould, Steven Stanley and others—"punctuationalists" confronting the facts of gaps and stasis in the fossil record and offering some novel nongradualist explanations—find themselves quoted by creationists to "establish" (much to their surprise) that the fossil record denies evolution (see Godfrey 1981). Punctuationalism has become a favorite target of creationist distortion and ridicule.

The last pages of Gish's book are devoted to ridiculing both modern punctuationalism and the return of Richard Goldschmidt's (1940) jerky or "saltational" evolution (see also Parker 1980; Morris 1975; Pratney 1982). Isn't punctuationalism, after all, the modern equivalent of Goldschmidt's "hopeful monster" model of evolutionary change? And isn't the return of Goldschmidt's "hopeful monster" a sure sign of desperation among evolutionists and of the impending collapse of evolutionary theory? Creationist Winkie Pratney sums up the argument this way:

> Recently, some like *Stephen Gould* of Harvard, have returned to the "hopeful monster" theory ("saltatory" (jumping) evolution, or the "punctuated equilibrium") of *Mr. Richard Goldschmidt* in the 1930s, the idea that *radical change* in genes or chromosomes made a lizard for instance, give birth to a bird, a "hopeful" idea indeed. Gould

> himself points out problems with this. (How VERY lucky can you get, and if you think *people* have problems finding a mate, how about our hopeful monster?) [1982, p. 8; italics in original]

In order to see how completely vacuous and absurd the creationist characterizations of both punctuationalism and Goldschmidt's "hopeful monster" metaphor really are, let us turn to each of these, and examine their empirical and theoretical foundations.

What Is Punctuationalism?

In 1972 Niles Eldredge of the American Museum of Natural History and Harvard paleontologist Stephen Jay Gould launched their new theory of evolution by "punctuated equilibria" in an attempt to explain the rarity, not absence, of transitional forms at the species level. Evolution, they said, commonly proceeds in fits and starts punctuating relatively long periods of evolutionary stasis. Ironically, Eldredge (1971) first arrived at his notion of punctuated equilibria while studying Devonian trilobites whose excellent fossil record includes intermediates at this fine level of evolutionary change. Gould (1969) similarly studied an excellent record of spatial and temporal variation in a Pleistocene land snail from Bermuda, *Poecilozonites.* It was the restricted temporal and geographic distribution of morphological intermediates and the pattern of occurrence of morphological innovations that convinced first Eldredge and then Gould that Darwin had been wrong in one important aspect of his theory of evolutionary change: his pervasive gradualism. The authors of the "theory of punctuated equilibria" raised Darwin's dilemma concerning the rarity of species-level transitions but offered a new solution—one foreshadowed by recent developments on quantum evolution (Simpson 1944, 1953), allopatric (or geographic) speciation (Mayr 1954), and rapid or "quantum" speciation (Lewis and Raven 1958; Lewis 1962, 1966; Carson 1959, 1971; Grant 1963, 1971). Eldredge and Gould posited a *biological* explanation of gaps in the fossil record, the strength of which lay in its ability to *simultaneously* account for two prominent features of the fossil record: (1) relatively long periods of stasis in some well-documented fossil species; and (2) intermittent interruptions of stasis by brief intervals of rapid evolution (often represented by gaps in the fossil record). Both stasis and rapid evolution, they reasoned, could be explained by the notion, already current in the neontological literature, that rapid morphological innovations occur in small "founder" populations often in conjunction with the evolution of reproductive barriers separating the founder populations from the parent populations. If successful, these innovations can lead to the growth and establishment of a "daughter" species or even to the eventual replacement of the mother species by the daughter species. This idea is decidedly non-Darwinian. That splitting of lineages (or "speciation") contributes significantly to morphological evolutionary change can be found nowhere in Darwin's work.

Darwin did, of course, offer a *biological* explanation for evolutionary stasis—an explanation that fit marvelously with his theory of descent with modification by natural selection. If evolutionary change occurs within populations by the (environmentally driven) slow accumulation of new beneficial variations, organisms that don't change *must* not change because their *environment* doesn't change. Punctuationalists find this kind of argument unsatisfactory (see Stanley 1981), but their main argument is with Darwin's nonbiological explanation for morphological gaps in the fossil record. Eldredge and Gould (1972) rejected as insufficient the notion that erosional hiatuses and depositional breaks explain these gaps. Here is their major source of disagreement with Darwin. Darwin (1859) had expected a *perfect* fossil record to document continuous sequences of morphological intermediates. Eldredge and Gould (1972) agreed that a perfect fossil record *would* document morphological intermediates between species, but they suggested that many of these would exhibit relatively brief and geographically limited existences. Indeed, Eldredge had such a near perfect record of the evolution of the Devonian trilobite *Phacops*. It was a record of stepwise evolutionary change in only two brief intervals during a span of eight million years! One such interval was recorded in a single easy-to-miss quarry in New York State. This quarry contained perfect intermediates between the geographically widespread mother and daughter species. In effect, due to the realities of an *imperfect* fossil record, most such intermediates will simply not be sampled.

Since the publication of Eldredge and Gould's seminal paper on punctuated equilibria, many paleontologists have tried to assess the prevalence of this pattern of evolutionary change in the organisms they know best (see especially Hallam 1977; Schwartz and Rollins 1979; Cracraft and Eldredge 1979, and numerous papers on the subject in the journal *Paleobiology*). Some have argued strongly in support of the theory (Stanley 1979, 1981; see also Williamson 1981 for one such case study); others have argued vehemently against it (Gingerich 1976; Cronin et al. 1981). In fact, the fossil record is often too coarse to choose between competing alternatives. Verne Grant made that point in 1977 when contrasting what he called "speciational" and "phyletic" trends. Grant had first proposed the concept of "quantum speciation" as an important process of macroevolution in 1963, and he then suggested that macroevolutionary trends could be products of consecutive quantum speciational events as well as products of gradual evolutionary change within lineages ("phyletic" trends). However, it will be difficult to tell those different kinds of trends apart "whenever the fossil record is incomplete, as it usually is. A speciational trend can easily be misread as a phyletic trend, and vice versa" (Grant 1977, p. 299). Thus it is quite possible that resolution of the problem of the validity of the Eldredge-Gould model as a general model of evolutionary change will come more from neontological than paleontological research.

In any case, it is obvious that intermediates exist at *both* fine and large scales in the fossil record. To the extent that evolution proceeds in a jerky manner at the species level, large-scale transitions that depend upon numerous fine-scale transitions will also be "herky jerky." And to the extent that speciation produces numerous closely allied forms, it is going to be difficult to identify direct ancestors precisely. Thus, speciation, because it is often "hidden" in the fossil record and because it increases the diversity of living organisms at any point in time, contributes to the difficulty of precisely identifying ancestral-descendant relationships at whatever scale.

Soon after punctuationalism had achieved recognition in biological and geological circles, creationists seized upon the opportunity to fashion it according to their own whims. "Stasis" must represent "variation within kinds." "Gaps" must represent "boundaries between kinds." Punctuationalists became scientists who had accurately deciphered the pattern of morphological variation in the fossil world but who, blinded by their evolutionary zeal, still refused to acknowledge the obvious: "Rather than forging links in the hypothetical evolutionary chain, the wealth of fossil data has only served to sharpen the boundaries between the created kinds" (Parker 1980, p. iii).

It is surprising that not even one creation scientist has objected to this gross distortion of the theory of "punctuated equilibria," because it makes a mockery of the concept of "created kind." Creationists allow "created kinds" to encompass lots of variability; entire genera, tribes, or even taxa of higher rank may represent "created kinds." "Kinds" are not "species," creationists insist, and they generally hold vehemently to that dogma except when describing the human "kind" (Marsh 1978). How, then, can creationists invoke "stasis" at the *species* level to explain "variation within kinds"? How can they use the data of punctuationalism as support for the concept of "created kind"? Do creationists mean to suggest that the God who created whole family-kinds also concerned himself with creating several different "kinds" of trilobites belonging to the *Phacops rana* species group, each with a slightly different number of vertical lenses in its compound eyes? Or that this God worried about paedomorphosis in the Bermuda land snail *Poecilozonites* enough to create four similar but discrete paedomorphic "kinds" within a single species group? Why are the boundaries between some *Phacops* "kinds" bridged by intermediate forms? Why do the paedomorphic land snails occur in such strata that they *appear* to have evolved independently from slightly different nonpaedomorphic parents? In fact, punctuationalism offers no support for creationism. Not in many hundreds of millions of years, much less the several thousand presumably available since a great flood reduced the variation of landbound kinds to that of single reproducing pairs, could "stasis" produce the variation creationists accept as variation "within kinds."

Logically, stasis should represent a supreme problem for the creationist scenario, which requires a phenomenal rate of splitting and evolutionary change in the last several thousand years to account for current levels of variation within supposed kinds! Unless creationists drastically alter their notion of "created kind," they should find no solace in a notion of stasis that applies only to the *species* (or, for fossils, "morphospecies") level of biological variation. And unless they drastically alter their argument that the gaps in the fossil record described by Eldredge, Gould, and other paleontologists reflect boundaries between "created kinds," they cannot defend a broad, indefinite (but clearly variable) concept of "created kind." (Creationists don't want to equate "kinds" with biological "species," because they know that speciation occurs. The way around this dilemma is to acknowledge its existence as a process of "variation within kinds.")

If creationists really wanted to accurately represent the empirical and theoretical foundations of punctuationalism, they would have to accept the fact that gaps in the fossil record are not boundaries between "created kinds." Then they would have to admit that punctuationalism offers *no* support whatever for the existence of "created kinds." If anything, the Eldredge-Gould model contradicts the prevalent creationist interpretation of processes of genetic change. Creationists maintain that change is produced gradually by natural selection acting on minor genetic variations among individuals reproducing "after their kind," but that such a process produces only variation within kinds. To those who have loudly protested that the concept of "kind" has been distorted and maligned by evolutionists who would mistake creationists' "kinds" for evolutionists' "species," punctuationalism must be seen as a competing theory of evolutionary change *within* "created kinds".

A Much-maligned Metaphor

Punctuationalism is not a theory of evolution by jumps but of evolution in spurts associated with speciation. But just how long is a "spurt"? Let us have the punctuationalists provide their own answer:

> New species usually arise, not by the slow and steady transformation of entire ancestral populations, but by the splitting off of small populations from an unaltered ancestral stock. The frequency and speed of such speciation is among the hottest topics in evolutionary theory today, but I think that most of my colleagues would advocate ranges of hundreds or thousands of years for the origin of most species by splitting. [Gould 1979, p. 26]

If a "spurt" lasts thousands of years, there is no need to postulate radical mechanisms to accomplish "rapid" evolutionary change, and, indeed, Eldredge and Gould (1972) offered none. Punctuationalism is not saltationism. Punctuationalism postulates the existence of morphological intermediates (as small populations), and in some of the best known cases (e.g., Eldredge 1971; Williamson 1981) such intermediates have been thoroughly documented. Saltationism postulates instead that gaps at the species level represent genetic revolutions achievable virtually instantaneously by mutations in the genetic blueprint for morphogenesis (development)—Goldschmidt's "systemic mutations." Thus, when Gould (1977*b*), in his column for *Natural History* magazine, talks about the "return of the hopeful monster," he is not talking about his theory of punctuated equilibria, but about developments that compliment his theory in explaining gaps in the fossil record at the species level.

Goldschmidt's notion of saltational evolution was quintessentially non-Darwinian, and its partial reemergence in the 1970s and early 1980s could not have been foreseen by those neo-Darwinists who sealed the coffin of saltationism with scorn back in the 1940s and 1950s. Why, in 1980, do we witness the publication of a collection of essays by leading geneticists favorably reassessing the scientific contributions of Richard Goldschmidt (Piternick 1980)? Is the reemergence of the "hopeful monster" a hopeless creation of evolutionists forging imaginary links between discrete kinds, as creationists would have us believe? Is it testimony to a growing desperation among people clinging to a forty-year-old fantasy in an effort to keep evolution alive? Or has the hopeful monster raised its ugly head again because somebody thinks, based on some real evidence, that Goldschmidt was at least partly right?

Richard Goldschmidt, 1878–1958, was a boy genius, a prolific writer, and a beloved university teacher (Stern 1980). But he was also the target of severe ridicule during his lifetime. He was eccentric; his ideas were outside the mainstream of genetics.

The son of a German-Jewish bourgeois family, Richard Goldschmidt was a voracious reader as a child, and he developed a lasting interest in natural science while still an adolescent, having even before then delved into archaeology, comparative linguistics, Goethe, and world literature. In his teens, he attained fluency in French, Italian, Latin, Greek, and English. At the University of Heidelberg he studied medicine briefly, then selected a career as a zoologist and completed a doctoral dissertation under the renowned zoologist Bütschli on aspects of the maturation and development of the trematode worm Polystomum. He then embarked upon various elaborate research projects and traveled periodically to Mediterranean marine laboratories, always compiling notes and posing questions. Before he settled on the field that would make (or, to some, break) his career, he had made meticulous contributions to the fields of histology, neurology, cytology, anatomy, and embryology. These early accomplishments earned him a place of prominence in his profession (see Stern 1980 for details).

Goldschmidt's chosen field was genetics, but he never stopped thinking about whole organisms, and as his study of chromosomes and hereditary transmission matured, so did his interest in developmental processes. As a developmental geneticist, he focused on how mutations affect developmental pathways and how such changes, in turn, affect evolution. This, he thought, was a major failing of population genetics, which concerned itself primarily with models of genetic transmission and not with genetic control of developmental processes. "It is regrettable that many evolutionists forget that the developmental potencies are a necessary partner of genetic change in the understanding of evolution," he wrote in 1955, three years before his death.

To Goldschmidt, saltations were abrupt changes in adult morphology, brought about by alterations in developmental processes and achievable within the confines of one or several mutational steps—the products of "macromutations." Macromutations are produced, Goldschmidt thought, by radical rearrangements of genetic material, through chromosomal inversions, translocations, or other aberrations. Conversant with natural variation in populations and with genetically controlled anomalies in numerous plants and animals, he rejected the notion that selection plus gradual accumulation of "micromutations" are sufficient to explain evolutionary change at all levels and invoked "macromutations" to explain species-level differences. When accused of developing a "cataclysmic" theory, he defended himself vigorously:

> There is nothing cataclysmic in a postulated process that occurs or is assumed to occur in a way that requires only known basic processes of chromosomal and genetic behavior. [1955, p. 491]

Rightly or wrongly, Goldschmidt sought genetic explanations for real (and sometimes quite successful) developmental anomalies that could form the basis of a theory of evolutionary change. If those anomalies normally help to initiate a sterility barrier, "radical" change in form could be a necessary by-product of a natural process by which new species are born.

In his copious writings, Goldschmidt displayed a fondness for metaphor. Population geneticists, he insisted, had figured out the "rules" of the "game" of evolution (the rules of selection), but they had not identified the "players" (micromutations or macromutations) (1948). The "hopeful monster" was merely another Goldschmidt metaphor—poorly chosen, perhaps, but nevertheless a *metaphor.* He did not intend to argue that reptiles gave birth to mammals or birds in one step!

Still, Goldschmidt's saltational view, even properly interpreted, must be recognized as extreme, and the existence of intermediate populations renders it unnecessary as an explanation of evolutionary change at the species level. Few evolutionists today would maintain that micromutations have *no* bearing on macroevolutionary change. But we are currently witnessing developments in evolutionary theory that can be read as confirmation of some of Goldschmidt's fundamental views. In fact, we know that speciation is achievable in small populations in only a few generations (White 1978; Gottlieb 1976; Carson 1971; Bush 1969, 1975, 1981; Lewis 1962, 1966, 1973; Lewis and Raven 1958). And we know that it can involve radical morphological change (Stanley 1979). The old orthodoxy that reproductive isolation is acquired very gradually through adaptive allelic substitution in geographically isolated large populations is being superseded by a new orthodoxy that poses numerous mechanisms, some quite radical, for achieving reproductive isolation.

Speciation also often involves rearrangements of the genetic material visible in the number and structure of the chromosomes—the organism's "karyotype." White concluded his book *Modes of Speciation* with the following words:

> One of the main conclusions of this book is that over 90 percent (and perhaps over 98 percent) of all speciation events are accompanied by karyotypic changes, and that in the majority of these cases the structural chromosomal rearrangements have played a primary role in initiating divergence. In general, however, population geneticists have neglected karyotypic changes and have tended to think of speciation in terms of the building-up of distinct gene pool of alleles without considering structural rearrangements of the chromosomes or looking upon such changes as simply equivalent to point mutations at individual loci. . . . Although the process of speciation is not exempt from the operations of the laws of population genetics, it almost certainly involves principles that do not operate in ordinary phyletic evolution.
> [1978, p. 324]

In addition to reevaluating mechanisms of speciation, morphologists, geneticists, and molecular biologists have also focused on the role of developmental regulation in producing morphological divergence (Britten and Davidson 1969, 1971; Davidson and Britten 1979; Valentine and Campbell 1975; King and Wilson 1975; Wilson, Carlson, White 1977; Bush 1975, 1981; Løvtrup 1974; Gould 1977*a*, 1980; Krone and Wolf 1977; Sarich 1980, Stanley 1979, 1981; Alberch 1980). It has become increasingly clear that small changes in the genes regulating development can have major consequences for adult morphology and that such changes can be rapidly fixed in small populations.

How can evolutionists fail to see Richard Goldschmidt's notion of the importance of genetic rearrangement and developmental regulation (as factors in speciation and rapid evolution) vindicated by these recent developments in biology? It is in this guise that the "hopeful monster," much subdued, has returned.

Richard B. Goldschmidt suffered many indignities during his own lifetime. He was sent to a civilian prison camp in Georgia in early 1918 for being German and detained there until after the First World War ended. (Several years earlier he had been prevented from returning to his German homeland after a brief research sojourn in Japan. The British blockade converted what was meant to be a stopover in San Francisco on his way home into a several-year-long residency, first in San Francisco and then in the eastern United States, where he worked in a laboratory at Yale University.) Later Goldschmidt was forced to seek permanent refuge in the United States by the growth of anti-Semitism and Naziism in his beloved Germany. Then, as a teacher at Berkeley, he suffered the ridicule of some of the great American population geneticists.

Now, many years after his death, Goldschmidt is being recognized as a great scientist (Piternick 1980). He asked the right questions. He posed daring answers. He was, in the words of Stern "a preceptor and critic of his era; designer of frameworks for the future" (1980, p. 88). To be sure, much of what Goldschmidt argued is now known to be wrong. He was wrong about the structure of DNA and the nature of the gene (Bakken 1980). He was wrong to reject, as he did, the notion that specific gene loci could be responsible for the production of specific enzymes. But he was right that chromosomes are more than strings of independent gene loci. In 1961, only three years after Goldschmidt died, Jacob and Monod produced experimental confirmation of the "operon model," demonstrating the existence of regulatory genes that control the activity of structural genes and giving substance to Goldschmidt's notion that the pattern or order of the genes within the chromosome is basic to their operation. Whether ordinary phyletic evolution can produce major changes in morphology is still the subject of much debate (Stebbins and Ayala 1981). Likewise the *dominance* of speciation as a factor in producing macroevolutionary change and the relative importance of chromosomal aberrations and of regulatory as opposed to structural gene changes in producing macroevolutionary shifts continue to be subjects of discussion and of argument.

But it is clear that major adaptational shifts can occur within a few generations and that speciation can occur in a geologic "instant." Whether by Goldschmidt's postulated mechanism, by "punctuated equilibria," or some other process, rapid evolution is known to take place.

At home I have a children's story that my son adores called *Harry and the Terrible Whatzit* (Gackenbach 1977). It is about a little boy who is afraid of big monsters, but who decides one day to confront the monster in his basement because his mother is down there and must be in danger. When the boy takes a good hard look at the monster, the monster retreats, begins to shrink, and finally, disappears. So it will be with the public perception of "hopeful monsters" as a menace to evolution. Take a good hard look and the menace disappears.

REFERENCES CITED

Alberch, Pere. 1980. Ontogenesis and morphological diversification. *American Zoologist* 120:653–67.

Bakken, Aimée H. 1980. Views on the nature of the gene, the structure and function of the chromosome, and heterochromatic heredity. In *Richard Goldschmidt: controversial geneticist and creative biologist,* ed. Leonie K. Piternick. *Experientia Supplementum* vol. 35, pp. 32–40. Basel/Boston/-Stuttgart: Birkhäuser Verlag.

Beck, Charles B. 1976. Origin and early evolution of angiosperms: a perspective. In *Origin and early evolution of angiosperms,* ed. Charles B. Beck, pp. 1–10. New York: Columbia Univ. Press.

Behrensmeyer, Anna D. and Hill, Andrew P., eds. 1980. *Fossils in the making: vertebrate taphonomy and paleoecology.* Chicago: Univ. of Chicago Press.

Britten, R. J. and Davidson, E. H. 1969. Gene regulation for higher cells: a theory. *Science* 165:349–57.

—. 1971. Repetitive and non-repetitive DNA sequences and a speculation on the origins of evolutionary novelty. *Quarterly Review of Biology* 46:111–38.

Bush, Guy L. 1969. Sympatric host race formation and speciation in Frugivorous Flies of the genus *Rhagoletis* (Diptera, Tephritidae). *Evolution* 23:237–51.

—. 1975. Modes of animal speciation. *Annual Review of Ecology and Systematics* 6:339–64.

—. 1981. Stasipatric speciation and rapid evolution in animals. In *Evolution and speciation,* eds. W. R. Atchley and D. S. Woodruff, pp. 201–18. Cambridge, Eng.: Cambridge Univ. Press.

Carson, Hampton L. 1959. Genetic conditions which promote or retard the

formation of species. *Cold Spring Harbor Symposia on Quantitative Biology* 24:87–105. Cold Spring Harbor, N. Y.: Biological Laboratory.
—. 1971. Speciation and the founder principle. *Stadler Symp.* 3:51–70.
Cloud, Preston. 1977. Scientific creationism: a new Inquisition brewing? *The Humanist,* vol. 37, no. 1, pp. 6–15.
Cracraft, Joel and Eldredge, Niles., eds. 1979. *Phylogenetic analysis and paleontology.* New York: Columbia Univ. Press.
Crompton, A. W. and Parker, Pamela. 1978. Evolution of the mammalian masticatory apparatus. *American Scientist* 66:192–201.
Cronin, J. E., Boaz, N. T., Stringer, C. B., and Rak, Y. 1981. Tempo and mode in hominid evolution. *Nature* 292:113–22.
Darwin, Charles. 1859. *The origin of species* (1963 ed.). New York: Washington Square Press.
—. 1879. Letter to J. D. Hooker. In *More letters of Charles Darwin: a record of his work in a series of hitherto unpublished letters,* vol. I. (1903), pp. 20–21. New York: Appleton.
Davidson, E. H. and Britten, R. J. 1979. Regulation of gene expression: possible role of repetitive sequences. *Science* 204:1052–59.
Doyle, J. A. and Hickey. L. J. 1976. Pollen and leaves from the mid-Cretaceous Potomac group and their bearing on early angiosperm evolution. In *Origin and early evolution of angiosperms,* ed. C. B. Beck, pp. 139–206. New York: Columbia Univ. Press.
Efremov, J. A. 1940. Taphonomy: a new branch of paleontology. *Pan-American Geologist,* vol. 74, no. 2, pp. 81–93.
Eldredge, Niles. 1971. The allopatric model and phylogeny in Paleozoic invertebrates. *Evolution* 25:156–67.
Eldredge, Niles and Gould, Stephen J. 1972. Punctuated equilibria: an alternative to phyletic gradualism. In *Models in Paleobiology,* ed. T. J. M. Schopf, pp. 82–115. San Francisco: Freeman, Cooper and Co.
Feduccia, J. Alan. 1980. *The age of birds.* Cambridge, Mass.: Harvard Univ. Press.
Gackenbach, Dick. 1977. *Harry and the terrible whatzit.* New York: Scholastic Book Services.
Gingerich, Philip. 1976. Paleontology and phylogeny: patterns of evolution at the species level in early tertiary mammals. *American Journal of Science* 276:1–28.
Gish, Duane. 1978. *Evolution? The fossils say no!* San Diego: Creation-Life Pubs.
Godfrey, Laurie R. 1981. The flood of antievolutionism. *Natural History,* (June), pp. 4–10.
Goldschmidt, Richard B. 1940. *The material basis of evolution.* New Haven, Conn.: Yale Univ. Press.
—. 1948. Ecotype, ecospecies and macroevolution. *Experientia* 4: 465–72.
—. 1955. *Theoretical genetics.* Berkeley and Los Angeles: Univ. of California Press.

Gottlieb, Leslie D. 1976. Biochemical consequences of speciation in plants. In *Molecular evolution*, ed. Francisco J. Ayala, pp. 123–40. Sunderland, Mass.: Sinauer Associates.

Gould, Stephen J. 1969. An evolutionary microcosm: Pleistocene and recent history of the land snail *P. (Poecilozonites)* in Bermuda. *Bulletin of the Museum of Comparative Zoology* 138:407–532.

—. 1977*a*. *Ontogeny and phylogeny*. Cambridge, Mass.: Harvard Univ. Press, Belknap Press.

—. 1977*b*. The return of the hopeful monsters. *Natural History* (June-July), pp. 22–30.

—. 1979. A quahog is a quahog. *Natural History* (August-September), pp. 18–26.

—. 1980. Is a new and general theory of evolution emerging? *Paleobiology* 6:119–30.

Gould, Stephen J. and Eldredge, Niles. 1977. Punctuated equilibria: the tempo and mode of evolution reconsidered. *Paleobiology* 3:115–51.

Grant, Verne. 1963. *The origin of adaptations*. New York: Columbia Univ. Press.

—. 1971. *Plant speciation*. New York: Columbia Univ. Press.

—. 1977. *Organismic evolution*. San Francisco: W. H. Freeman & Co.

Hallam, A., ed. 1977. *Patterns of evolution as illustrated by the fossil record (developments in palaeontology and stratigraphy, 5.)* Amsterdam/Oxford /New York: Elsevier Scientific Pub. Co.

Harris, T. M. 1960. The origin of angiosperms. *Advancement of Science* 17: 207–13.

Hildebrand, Milton. 1974. *Analysis of vertebrate structure*. New York: Wiley.

Hinchliffe, J. R. and Johnson, D. R. 1980. *The development of the vertebrate limb: an approach through experiment, genetics and evolution*. Oxford, Eng.: Clarendon Press.

Hughes, Norman F. 1976. *Palaeobiology of angiosperm origins: problems of Mesozoic seed-plant evolution*. Cambridge, Eng.: Cambridge Univ. Press.

Jacob, F. and Monod, J. 1961. Genetic regulatory mechanisms in the synthesis of proteins. *Journal of Molecular Biology* 3:318–56.

Jenkins, F. A., Jr. and Parrington, F. R. 1976. The postcranial skeletons of the Triassic mammals *Eozostrodon, Megazostrodon* and *Erythrotherium*. *Philosophical Transactions of the Royal Society of London*, Series B, vol. 273, pp. 387–431.

Johanson, Donald C. and Edey, Maitland. 1981. *Lucy: the beginnings of humankind*. New York: Simon and Schuster.

King, Marie-Claire and Wilson, Allan C. 1975. Evolution at two levels: molecular similarities and biological differences between humans and chimpanzees. *Science* 188:107–16.

Krone, W. and Wolf, V. 1977. Chromosome variation and gene action. *Hereditas* 86:31–36.

Lanham, Url. 1973. *The bone hunters*. New York: Columbia Univ. Press.

Lewis, Harlan. 1962. Catastrophic selection as a factor in speciation. *Evolution* 16:257–71.
—. 1966. Speciation in flowering plants. *Science* 152:167–72.
—. 1973. The origin of diploid neospecies in *Clarkia. American Naturalist* 107:161–70.
Lewis, Harlan and Raven, Peter H. 1958. Rapid evolution in *Clarkia. Evolution* 12:319–36.
Lillegraven, Jason A., Kielan-Jaworowska, Zofia, and Clemens, W. A., eds. 1979. *Mesozoic mammals: the first two-thirds of mammalian history.* Berkeley and Los Angeles: Univ. of California Press.
Løvtrup, Soren. 1974. *Epigenetics: a treatise on theoretical biology.* New York: Wiley.
Luckett, W. Patrick and Szalay, Frederick S., eds. 1975. *Phylogeny of the primates: a multidisciplinary approach.* New York: Plenum Press.
Marsh, Frank Lewis. 1978. Variation and fixity among living things: a new biological principle. *Creation Research Society Quarterly* 15:115–18.
Mayr, Ernst. 1954. Change of genetic environment and evolution. In *Evolution as a process,* eds. J. Huxley, A. C. Hardy, and E. B. Ford, pp. 157–80. London: Allen and Unwin.
Morbeck, Mary Ellen, Preuschoft, Holger, and Gomberg, Neil. 1979. *Environment, behavior, and morphology: dynamic interactions in primates.* New York: Gustav Fischer.
Morris, Henry M. 1975. *The troubled waters of evolution.* San Diego: Creation-Life Pubs.
Ostrom, John H. 1975. The origin of birds. *Annual Review of Earth and Planetary Sciences* 3:55–77.
—. 1976. *Archaeopteryx* and the origin of birds. *Biological Journal of the Linnean Society* 8:91–182.
—. 1979. Bird flight: how did it begin? *American Scientist* 67:46–56.
Panchen, A. L., ed. 1980. *The terrestrial environment and the origin of land vertebrates.* Systematic Assoc. Special Vol. No. 15. New York: Academic Press.
Parker, Gary E. 1980. Creation, selection and variation. ICR Impact Series, no. 88, pp. i–iv.
Piternick, Leonie K., ed. 1980. *Richard Goldschmidt: controversial geneticist and creative biologist. Experientia Supplementum* vol. 35. Basel/Boston/-Stuttgart: Birkhäuser Verlag.
Pratney, Winkie. 1982. Creation or evolution? Part III. The fossil record. *The Last Days Newsletter,* vol. 5, no. 1, pp. 7–11, 32–34. Lindale, Tex.: Last Days Ministries.
Sarich, Vincent M. 1980. A macromolecular perspective on the material basis of evolution. In *Richard Goldschmidt: controversial geneticist and creative biologist,* ed. Leonie K. Piternick, pp. 27–31. *Experientia Supplementum* vol. 35. Basel/Boston/Stuttgart: Birkhäuser Verlag.
Schultze, Hans-Peter. 1977. The origin of the tetrapod limb within the Rhipidis-

tian fishes. In *Major patterns of vertebrate evolution,* eds. Max Hecht, P. C. Goody, and B. M. Hecht, pp. 541–544. New York: Plenum Press.

Schwartz, Jeffrey H. and Rollins, Harold B. 1979. *Models and methodologies in evolutionary theory. Bulletin of Carnegie Museum of Natural History,* no. 13. Pittsburgh, Pa.

Shipman, Pat. 1981. *Life history of a fossil: an introduction to taphonomy and paleoecology.* Cambridge, Mass.: Harvard Univ. Press.

Simpson, George Gaylord. 1944. *Tempo and mode in evolution.* New York: Columbia Univ. Press.

—. 1953. *The major features of evolution.* New York: Columbia Univ. Press.

Slaughter, Bob H. 1970. Evolutionary trends of Chiropteran dentitions. In *About bats: a Chiropteran biology symposium,* eds. Bob H. Slaughter and Dan W. Walton, pp. 51–83. Dallas, Tex.: Southern Methodist Univ. Press.

Stanley, Steven M. 1979. *Macroevolution: pattern and process.* San Francisco: W. H. Freeman & Co.

—. 1981. *The new evolutionary timetable: fossils, genes and the origin of species.* New York: Basic Books.

Stebbins, G. Ledyard and Ayala, Francisco J. 1981. Is a new evolutionary synthesis necessary? *Science* 213:967–71.

Stern, Curt. 1980. Richard Benedict Goldschmidt (1878–1958): a biographical memoir. In *Richard Goldschmidt: controversial geneticist and creative biologist,* ed. Leonie Piternick, pp. 68–99, *Experientia Supplementum* vol. 35. Basel/Boston/Stuttgart: Birkhäuser Verlag. (Originally published in 1967 as a memoir of the National Academy of Sciences.)

Szalay, Frederick S. and Delson, Eric. 1979. *Evolutionary history of the primates.* New York: Academic Press.

Szarski, Henry K. 1977. Sarcopterygii and the origin of tetrapods. In *Major patterns of vertebrate evolution,* eds. Max K. Hecht, Peter C. Goody, and B. M. Hecht, pp. 517–40. New York: Plenum Press.

Valentine, James W. 1977. General patterns of Metazoan evolution. In *Patterns of evolution as illustrated by the fossil record,* ed. A. Hallam. pp. 27–57. Amsterdam: Elsevier.

—. In press. The origin of complex organisms. In *A century after Darwin: current issues in evolution,* ed. L. Godfrey. Boston: Allyn & Bacon.

Valentine, James W. and Campbell, Cathryn A. 1975. Genetic regulation and the fossil record. *American Scientist* 63:673–80.

White, Michael J. D. 1978. *Modes of speciation.* San Francisco: W. H. Freeman and Co.

Williamson, P. G. 1981. Palaeontological documentation of speciation in Cenozoic molluscs from Turkana basin. *Nature* 293:437–43.

Wilson, A. C., Carlson, S. S. and White, T. J. 1977. Biochemical evolution. *Annual Review of Biochemistry* 46:573–639.

12

Fossils, Stratigraphy, and Evolution: Consideration of a Creationist Argument

BY STEVEN D. SCHAFERSMAN

Introduction

Scientists today are embroiled in a controversy involving creation and evolution. Evolution, a fact of science, has long been challenged by the alternative theistic and supernaturalistic hypothesis of creation, a doctrine espoused by a number of fundamentalist religions. Although this controversy dates back to the beginning of the nineteenth century, it appears today in a new guise, that of creationism masquerading as science. "Scientific creationism," as it is called, was created in the 1960s as a result of the repeated failures of creationists in state legislatures and courts of law to win approval of their desire to teach biblical creationism in public school science classes. Both constitutional law and court decisions have continually prohibited teaching religious doctrines in public schools, so scientific creationism was born to try to sneak creationism, as "science," into the classroom.

This chapter will examine a single argument of the "scientific" creationists: that geologists use fossils to date rocks and rocks to date fossils and that this is circular reasoning because it assumes the truth of evolution (by using a fossil sequence of simple to complex to date rocks) while it purports to provide evidence for the existence of evolution (by using the rock sequence thus constructed to demonstrate an evolutionary sequence of simple to complex organisms through time). In this chapter I will examine and refute only this single creationist argument so that I can reveal in detail the extent to which the creationists use sophistic reasoning, quote scientists out of context, and distort scientific knowledge to make their invalid arguments seem plausible. I would not be able to provide such detailed documentation of these practices in the space available were I to examine a number of different arguments. But I do claim that the same documented refutation can be made for any creationist argument.

The Creationist Style of Argument

Louis Agassiz, one of the last great scientific creationists, wisely admonished his fellow scientists to "study nature, not books." The modern creationists, however, study books, not nature. Specifically, they study the Bible and the books of scientists; their research effort is devoted to making their beliefs about the empirical world consistent with the former and to searching the latter for appropriate quotations to use (or misuse) in their campaign against science and evolution. Creationist tracts and books never scientifically examine the nature of the creation in which they believe, for all they can state about creation and its results are fiat assertions and personal speculations. Instead, they claim that if evolution is false, then creation is true, and then try to demonstrate that evolution is false.

Scientists rarely analyze creationist publications for the purpose of correcting the errors and enlightening the public because of the time such activity takes away from their research, because of the necessity of being familiar with creationist arguments as well as science, and mainly because to answer every argument and correct each distortion would require the scientist to write much more and explain things in much greater detail than the creationist. The creationist relies to a large extent on the sheer volume of distortions, facile arguments, and superficial explanations to achieve the desired end of confusing the reader and on suggesting antiscientific implications that remain in the reader's memory, rather than on persuading the reader with convincing arguments and evidence. Lay readers are particularly vulnerable to this technique for they generally have an inadequate reference base to evaluate the claims, even if they are skeptical.

The Alleged Circularity of Biostratigraphy

The claim that the methods of biostratigraphy are circular is one of the most frequently used creationist arguments. Henry M. Morris, director of the Institute for Creation Research, is the primary exponent of this argument, but other creationists such as Gish (1973, p. 39), Wysong (1976, pp. 352–54), Wilder-Smith (1980, p. 103), and Parker (1980, p. 123) have also popularized it. All of these creationists hold that only "by prior commitment to evolutionary theory" or by "the assumption of evolution" is it possible to arrange fossil-bearing rocks "in a supposed time-sequence known as the geologic column." Wilder-Smith states that even "a little reflection shows that a prerequisite for [the index fossil method of dating's] reliability is the reliability of the entire Neodarwinian theory of evolution. . . . In reality, the index fossil method serves as a classic example of circular thought, for it accepts the assumptions of evolutionary theory . . . in order to determine the age of man and other organisms" (1980, p. 103). Similar statements can be found in many other creationist publications.

But let us first turn to Henry Morris's definitive statement of this argument:

> Creationists have long insisted that the main evidence for evolution—the fossil record—involves a serious case of circular reasoning. That is, the fossil evidence that life has evolved from simple to complex forms over the geological ages depends on the geological ages of the specific rocks in which these fossils are found. The rocks, however, are assigned geologic ages based on the fossil assemblages which they contain. The fossils, in turn, are arranged on the basis of their assumed evolutionary relationships. Thus the main evidence for evolution is based on the assumption of evolution. [Morris 1977, p. i]

After this opening statement, Morris quotes and paraphrases scientists to support his argument. First he lets O'Rourke speak:

> These principles have been applied in [biostratigraphy], which starts from a chronology of index fossils, and imposes them on the rocks. Each taxon represents a definite time unit and so provides an accurate, even "infallible" date. If you doubt it, bring in a suite of good index fossils, and the specialist without asking where or in what order they were collected, will lay them out on the table in chronological order.
> [O'Rourke 1976, pp. 51–52, as cited by Morris 1977, p. i]

Then he explains:

> That is, since evolution always proceeds in the same way all over the world at the same time, index fossils representing a given stage of evolution are assumed to constitute infallible indicators of the geologic age in which they are found. This makes good sense and would obviously be the best way to determine relative geologic age—if, that is, we knew infallibly that evolution were true! [Morris 1977, p. i]

O'Rourke's (1976) paper was published in the *American Journal of Science,* a respectable geological journal. The paper is, nevertheless, full of errors of fact and reasoning about biostratigraphy, and it should not have escaped the criticism of the journal's reviewers. But it did and thus Morris uses it for all it is worth. Because of the source, each of O'Rourke's statements seems to have the authority of science behind it. As far as I know, no geologist has ever replied to O'Rourke's paper, possibly because no one felt the need in the face of such grossly mistaken notions. (The same phenomenon is reflected in the general refusal of geologists to reply to lengthy tracts against plate tectonics and continental drift published in authentic scientific journals by the few legitimate geologists who do not accept the new global tectonic theory.) For the present, therefore, this chapter must serve as a reply to O'Rourke as well as to Morris.

O'Rourke's description of the principles of biostratigraphy is simply inaccurate. Biostratigraphy does not start from a chronology of index fossils and impose them on the rocks. Rather, the relative sequence of index fossils is determined from the superposition of horizontal strata; that is, from the relative sequence of horizontally layered fossil-bearing sedimentary rocks. Biostratigraphy rests on three laws or principles of stratigraphy that are universally accepted. First, *sedimentary strata are initially deposited horizontally.* This "Law of Initial Horizontality" is based on empirical observation, laws of hydrodynamics, and the scientific working hypothesis of *actualism,* which states that natural laws, such as the laws of physics, are unchanging or uniform through time. That is, we assume that sediments in the geologic past were invariably deposited horizontally, just as we universally observe them to be today. Second, *younger undisturbed strata invariably overlie older undisturbed strata.* This "Law of Superposition" is also based on empirical observation, physical law (gravity), and actualism. We have yet to observe older sediments being deposited on top of younger sediments, and for this to occur numerous physical laws would have to be violated. (We will discuss the phenomenon of disturbance of strata later.) Applying the first and second principles of stratigraphy, we can describe the sequence of fossils at any one locality of undisturbed strata and thus immediately possess a preliminary relative chronology of index fossils. This chronology is derived *from* the rocks, not "imposed" upon them as O'Rourke maintains. The third principle or law allows us to extend our biochronology to other areas. This is the "Law of Biotic Succession," which states that the *fossils in the strata will always occur in the same sequence, regardless of geographic location.* It is important to note that evolution has nothing to do with this principle; rather, it is based on empirical observation and actualism. Never has a single fossil been found out of stratigraphic sequence (of course, reworking and disturbance of strata by thrusting or overturning of beds do cause fossils to be found out of their proper vertical sequence, but these are well-understood phenomena that can be rationally explained and do not legitimately enter into our theoretical discussion). Because the sequence of fossil succession is invariable, geologists are able to infer that rocks with the same fossils are the same age. This was first confirmed by laterally tracing fossil-bearing beds wherever possible. Geologists discovered that no matter how far they traced the strata, *both* the reoccurring sequence of fossils and the position of fossils within identical rock strata of the same vertical relationship remained constant.

In practice, the sequence of fossils at one locality of undisturbed strata, designated as a *type section,* is measured and described. This step makes use of the principles of initial horizontality and superposition; the end result is a sequence—an *ordinal* time scale (usually referred to as a *relative* time scale)—that utilizes fossils. Subsequently, correlations are made and improved by interpolating more and better sections back to the type section. This activity makes use of the principle of biotic succession; the result is a much refined ordinal time scale that can be used on a global basis. This *biostratagraphic* time scale (based on the sequence of fossils) is combined with *lithostratigraphy* ("litho" meaning rock, the sequence of rocks) to construct a *chronostratigraphic* time scale that ideally allows one to date any rock on earth to its precise age on the geochronologic time scale (which is simply a *global* chronostratigraphic time scale). Today it is possible, by the use of radiometric dating, to assign a number of years to the geochronologic time scale; this results in an *interval* time scale (usually and unfortunately called an *absolute* time scale). The most important thing to note here, and indeed the most important point in this whole chapter, is that evolution plays absolutely no role in the principles of stratigraphy. Indeed, evolution is completely irrelevant to biostratigraphy. The basis of biostratigraphy is *litho*stratigraphy—that is, the construction of a biostratigraphic time scale from the observed sequence of material objects in the sedimentary rock record. Therefore, geologists do *not* assume any evolutionary relationships when using biostratigraphy to determine the ages of rocks that contain fossils.

If we return to the earlier quotations, we now can understand that O'Rourke is wrong, but Morris paraphrases O'Rourke's quote into something even more incorrect by adding things that O'Rourke never even suggested, such as "index fossils representing a given stage of evolution are assumed to constitute infallible indicators of the geologic age . . ." (1977, p. i). O'Rourke said nothing about "stage of evolution," but Morris clearly wants his readers to think he did. Morris had previously claimed that fossils "are arranged on the basis of their assumed evolutionary relationships," which to Morris means arranged in an order of simple to complex. If one actually reads O'Rourke's original paper, it is obvious that O'Rourke is opposed to Morris's thesis. O'Rourke (1976, p. 52) explicitly disagrees with Schindewolf (1960, p. 18), who had maintained that "time units are derived from . . . the development of organisms. . . ." To this O'Rourke replies that "before [evolution by natural selection] could be used to mark off time units, it had to be calibrated by comparing it to the geologic column, which, as a matter of fact, was put together before evolution was known" (1976, p. 52). O'Rourke's criticism of

Schindewolf (and therefore of the same argument by Morris) is completely valid, and Morris is clearly misinterpreting O'Rourke's argument concerning "circularity" in biostratigraphic reasoning.

Morris next quotes Gareth Nelson, a systematist:

> That a known fossil or recent species, or higher taxonomic group, however primitive it might appear, is an actual ancestor of some other species or group, is an assumption scientifically unjustifiable, for science never can simply assume that which it has the responsibility to demonstrate. It is the burden of each of us to demonstrate the reasonableness of any hypothesis we might care to erect about ancestral conditions, keeping in mind that we have no ancestors alive today, that in all probability such ancestors have been dead for many tens or millions of years, and that even in the fossil record they are not accessible to us.
>
> [Nelson 1969, p. 27, as cited by Morris 1977, pp. i–ii]

Morris goes on to interpret Nelson's argument:

> There is, therefore, really no way of proving scientifically any assumed evolutionary phylogeny, as far as the fossil record is concerned.
>
> [Morris 1977, p. ii]

Here we have a case of gross distortion. Morris is putting words in Nelson's mouth. Nelson said that we could not assume an ancestral-descendant relationship, that it had to be *demonstrated,* and that to do this is very difficult. He did not say that there was "really no way" of doing this, and he certainly did not say that there was no way of "proving" such a relationship. It is an elementary fact that science never "proves" anything; proof is found only in such disciplines as logic or mathematics where initial conditions are defined. *Any* knowledge system based on empiricism must falsify or corroborate hypotheses to determine truth, not "prove" something to achieve it. Therefore, Morris's frequent demand for "proof" is an extremely common creationist tactic to throw scientists off balance in the public's eyes, since the public has the mistaken impression that science *does* prove things, and scientists, when asked for proof, cannot provide it. Again, Morris relies on the public's poor knowledge of science to achieve his goal of discrediting science and evolution.

> It would help if the fossil record would yield somewhere at least a few transitional sequences demonstrating the evolution of some kind of organism into some other more complex kind. So far, however, it has been uncooperative. [Morris 1977, p. ii]

> The abrupt appearance of higher taxa in the fossil record has been a perennial puzzle. If we read the record rather literally, it implies that organisms of new grades of complexity arose and radiated relatively rapidly.
> [Valentine and Campbell 1975, p. 673, as cited by Morris 1977, p. ii]

Here Morris uses a quotation from a pair of scientists, which accurately describes one aspect of the fossil record, to support his invalid claim that there are no transitional sequences in the fossil record of organisms that evolve into new grades of complexity. First, Morris's claim is untrue: there are a number of well-documented transitional sequences of less complex to more complex grades of organisms, such as all the transitions of the vertebrate classes. Second, Valentine and Campbell are making a statement about the abrupt appearance of "higher taxa" of greater complexity in the fossil record. These higher taxa, usually orders and families in the case of vertebrates, are the groups or elements that actually constitute the well-documented transitions. The abrupt appearance of elements *within* a transition, caused by a poor fossil record or punctuated evolution, does not negate the existence of the transition. Morris wishes to imply, deceptively and illogically, that it does.

But at this point in his argument, Morris has only begun his attempt to discredit biostratigraphy. His technique is to gather every possible quotation he can find, whether relevant to his thesis or not, that contains the words "circular," "circularity," or "circular reasoning."

> The dating of the rocks depends on the evolutionary sequence of the fossils, but the evolutionary interpretation of the fossils depends on the dating of the rocks. No wonder the evolutionary system, to outsiders, implies circular reasoning. [Morris 1977, p. ii]

> The intelligent layman has long suspected circular reasoning in the use of rocks to date fossils and fossils to date rocks. The geologist has never bothered to think of a good reply, feeling the explanations are not

> worth the trouble as long as the work brings results. This is supposed to be hard-headed pragmatism.
>
> [O'Rourke 1976, p. 47, as cited by Morris 1977, p. ii]

Morris again quotes O'Rourke to support his argument of circular reasoning, but as previously explained, O'Rourke's judgment is in error. The geologist has never bothered to think of a good reply because none is necessary: layman suspicion of circular reasoning does not mean that geologists use such reasoning. The possibility of layman ignorance should be investigated before geologists are charged with using illogical methods. Morris continues:

> The main 'result' of this system [of dating], however, is merely the widespread acceptance of evolution. It is extremely inefficient in locating oil or other economically useful deposits. Perhaps, however, geologists feel that, since biologists had already proved evolution, they are justified in assuming it in their own work. But biologists in turn have simply assumed evolution to be true. [Morris 1977, p. ii]

Morris quotes philosopher Kitts:

> But the danger of circularity is still present. For most biologists the strongest reason for accepting the evolutionary hypothesis is their acceptance of some theory that entails it. There is another difficulty. The temporal ordering of biological events beyond the local section may critically involve paleontological correlation, which necessarily presupposed the non-repeatability of organic events in geologic history. There are various justifications for this assumption but for almost all contemporary paleontologists it rests upon the acceptance of the evolutionary hypothesis.
>
> [Kitts 1974, p. 466, as cited by Morris 1977, p. ii]

On one occasion in Houston I personally heard Morris argue that biostratigraphy involves circular reasoning and then belittle the many hardworking biostratigraphers employed in the oil industry by stating that their discipline was "extremely inefficient in locating oil" and that the fact that only 10 percent of all new wells drilled were successful was evidence for this. Of all the distortions and illogical claims Morris made in that lecture, this was the only one that set my teeth on edge, because it was an insult to men and women whose work is absolutely vital to the oil industry. Can anyone seriously believe that the oil industry would employ hundreds of highly paid professionals whose work was "extremely inefficient in locating oil"? Morris's statistical citation of a success rate of only 10 percent is correct, but he is apparently ignorant of the many real reasons why the great majority of new wells are failures. Biostratigraphers, of course, do not assume that biologists have "proved" evolution (as if this were possible). They universally accept evolution because they are scientists, but even this is irrelevant, since they never use evolution in their work.

Morris shrewdly cites Kitts to justify his claim. Kitts is a philosopher of science whose published work is often better informed than the above quotation would suggest. In that quotation, Kitts betrays an ignorance of certain aspects of both philosophy and science. First, biologists do not accept evolution because of their acceptance of some theory that entails it, but because spontaneous generation of complex organisms has been shown not to occur and because evolution is the only materialistic alternative. They also accept evolution because the evidence for it is very strong; recognition of this fact does not require acceptance of any specific theory or process (see Gould 1981). Second, Kitts has accepted the creationists' myth that acceptance of some aspect of evolution is necessary to perform biostratigraphy, the "ordering of biological events beyond the local section [by] correlation," As I have shown, this is entirely false. Third, Kitts implies that acceptance of evolution justifies the assumption of nonrepeatability of organic events. This is erroneous; evolution does not assume or require nonrepeatability, and modern evolutionary theory certainly allows it. Fourth, Kitts asserts that biostratigraphy "necessarily" presupposes the nonrepeatability of organic events. This is false because repeated events would not necessarily make biostratigraphy unworkable; in fact, there exist documented examples of repeated or "iterative" evolution of homeomorphs from the same and different lineages. These cases occur, for example, in the planktonic foraminifera: the repeated homeomorphs are said to be absolutely indistinguishable except by position in the stratigraphic record; these species are nevertheless used in biostratigraphy in concurrent range zones, where two overlapping species are needed to establish the presence of a biozone (biozones are the "units" of biostratigraphy). Thus, the assumption of nonrepeatability is not needed in either theory or practice for biostratigraphic correlation to work.

Nonrepeatability is only assumed for complex sequences in the construction of evolutionary relationships and histories (*genealogies* and *phylogenies*).

Morris next turns to a series of authors whose true meanings he must distort or hide in order to make his case. First in line is Dr. Derek Ager, former president of the British Geological Association, who Morris presents as one of several modern geologists "now recognizing the existence of circular reasoning in their geological methodologies" (1977):

> It is a problem not easily solved by the classic methods of stratigraphical paleontology, as obviously we will land ourselves immediately in an impossible circular argument if we say, firstly that a particular lithology is synchronous on the evidence of its fossils, and secondly that the fossils are synchronous on the evidence of the lithology.
>
> [Ager 1973, p. 62, as cited by Morris 1977, p. ii]

Creationists like to quote authorities who are distinguished by some professional office or position. This gives the supporting quotation greater credibility, although the creationist may present the same authority in the next paragraph as an ignorant obscurantist who believes in the logic and reliability of science despite the "evidence" to the contrary. Morris hides the fact that Ager is speaking of the reliability of chronostratigraphy (the determination of the exact *age* of strata) and not of biostratigraphy (the determination of the *sequence* of strata), and Ager's comments are perfectly correct in this context. Note that Ager uses the word "synchronous," which means "occurring at the same time." The attempted determination of synchroneity is a goal of chronostratigraphy, and the rock-fossil argument would be circular if used as Ager describes. Again, Morris wishes to reinforce the theme of "circularity" by repeating the word, no matter how irrelevant the context.

Morris now introduces Ronald West as "another geologist who has recognized the circularity problem":

> Contrary to what most scientists write, the fossil record does not support the Darwinian theory of evolution because it is this theory (there are several) which we use to interpret the fossil record. By doing so, we are guilty of circular reasoning if we then say the fossil record supports this theory.
>
> [West 1968, p. 45, as cited by Morris 1977, p. iii]

Careful reading of this quotation will show that the argument actually being made is totally irrelevant to Morris's thesis. West is correctly asserting that the fossil record supports the *fact* of evolution, not a particular *theory* of evolution. He is not talking about any alleged circularity in biostratigraphy at all. Here Morris is using a statement about alternative evolutionary explanations of the fossil record as an argument against the actual occurrence of evolution and simultaneously misleading his readers into thinking that West's argument is about circularity in the methods of biostratigraphy.

Morris next turns to a technical paper on paleontological methods, commenting that this paper is yet another acknowledgment of "the circular reasoning process involved in developing paleontological sequences" (Morris 1977, p. iii).

> The prime difficulty with the use of presumed ancestral-descendant sequences to express phylogeny is that biostratigraphic data are often used in conjunction with morphology in the initial evaluation of relationships, which leads to obvious circularity.
>
> [Schaeffer, Hecht, Eldredge 1972, p. 39, as cited by Morris 1977, p. iii]

> In view of such admissions from many leading evolutionists, it is clear that there neither is, nor can be, any *proof* of evolution. The evidence for evolution is merely the assumption of evolution.
>
> [Morris 1977, p. iii]

This quote of Schaeffer et al. is so taken out of context that it is difficult for a trained paleontologist, much less an average reader, to understand the true context of the remark without referring back to the original manuscript. Nevertheless, it is obvious, with reflection, that the quote does not support Morris's argument of circularity in biostratigraphy. Not only are ancestral-descendant sequences unnecessary for biostratigraphy, but the authors are clearly referring to something other than biostratigraphy as circular. Schaeffer et al. are objecting to the use of stratigraphic data to *recognize* ancestral-descendant sequences, because biostratigraphic data are commonly used to *evaluate* such sequences. They maintain that only comparative morphology (not stratigraphic sequence) should be used to reconstruct evolutionary history. This phylogenetic sequence can then be tested, in part, by reference to a stratigraphically derived sequence of fossils. Once again, Morris's supporting quote is irrelevant to his argument, and Morris is including it either because he does not understand the quote or because he wants to deceive his readers. Furthermore, he draws the absurd conclusion that all of the previous quotations are clear evidence that there is no "*proof* of evolution." Not a single quotation in his article justifies this conclusion, which is true only in the philosophical sense that "proof" is limited to logical, not empirical, knowledge systems.

Morris concludes his article by presenting a great number of quotations from O'Rourke with accompanying deceptive preambles. Three are considered below.

> The fiction that the geological column was actually represented by real rock units in the field has long been abandoned, of course.
> [Morris 1977, p. iii]

> By mid-nineteenth century, the notion of 'universal' rock units had been dropped, but some stratigraphers still imagine a kind of global biozone as 'time units' that are supposed to be ubiquitous.
> [O'Rourke 1976, p. 50, as cited by Morris 1977, p. iii]

The creationists notoriously claim that the geologic column (geochronologic time scale) is a fiction, because there is no place in the world where you can see it. O'Rourke correctly mentions that nineteenth-century geologists discarded the notion of "universal" rock units, i.e., the same rock units deposited universally over the earth. There are no such things. But what Morris claims in his distortion of O'Rourke's quotation is false. Of course the geologic column is represented by "real rock units in the field." Although not universal in extent, such rock units, visible in any highway roadcut, are representatives of the geologic column.

> And if physical data in the field seem in any case to contradict this assumed evolutionary development, then the field data can easily be reinterpreted to correspond to evolution! This is always possible in circular reasoning. [Morris 1977, p. iii]

> Structure, metamorphism, sedimentary reworking and other complications have to be considered. Radiometric dating would not have

> been feasible if the geologic column had not been erected first. The axiom that no process can measure itself means that there is no absolute time, but this relic of the traditional mechanics persists in the common distinction between "relative" and "absolute" age.
> [O'Rourke 1976, p. 54, as cited by Morris 1977, pp. iii–iv]

Morris's complete misrepresentation of this quotation from O'Rourke is so flagrant that no comment is necessary. O'Rourke, however, presents a confused notion of the geologic concept of "absolute time," which is simply the expression of age on an interval scale rather than on the ordinal scale of biostratigraphy and is not meant to be absolute in the popular sense O'Rourke is construing it to be. Also, radiometric dating is completely independent of the geologic column, which was constructed solely by biostratigraphy and lithostratigraphy, so its feasibility is definitely *not* dependent on the historically prior existence of the geologic column. In fact, the extremely high degree of congruence between the independent radiometric and chronostratigraphic time scales is excellent corroborating evidence for the veracity of both.

> He does recognize, however, that if the actual physical geological column is going to be used as a time scale, it is impossible to avoid circular reasoning. [Morris 1977, p. iv]

> The rocks do date the fossils, but the fossils date the rocks more accurately. Stratigraphy cannot avoid this kind of reasoning if it insists on using only temporal concepts, because circularity is inherent in the derivation of time scales.
> [O'Rourke 1976, p. 53, as cited by Morris 1977, p. iv]

This last quotation requires a reply to O'Rourke. It is impossible for fossils to date rocks more accurately than rocks date fossils, for fossil dating is based entirely on the relative dates (sequence) of rocks. The rocks date the rocks by means of the fossils, and there are no exceptions. Furthermore, circularity is not inherent in the derivation of time scales; I think any scientist would be surprised to hear this. Rather, time scales can best be described as arbitrary, since the unit of time measurement, the moment of time zero, and the direction of time's flow are all arbitrary decisions.

Another Creationist Treatment of Rocks and Fossils

We turn now to the most important creation-science textbook written for public school consumption—*Scientific Creationism* (Morris 1974). Unlike most creation-science literature, this book actually lists "predictions" that the creationists claim they can make for the existence of a "Creator" and for the special creation of the universe, earth, life, and species. For example, it claims that the "creation model" predicts the first and second laws of thermodynamics, the constancy of natural law, the existence of intelligence in man, gaps in the fossil record, and so forth. Clearly these "predictions" are for phenomena that are either self-evident or readily acceptable to the average layman. But they are not predictions of creationism at all because their relationship to a "Creator" (or any supernatural phenomenon) cannot be tested. Genuine scientific predictions are generated by a theory through cause and effect relationships. The path from theory to prediction should be defined and the mechanical processes and material products specified; without this, predictions cannot be tested by empirical methods. The creationists have never tried to develop such a theory of creation; indeed, they insist it cannot be done. The scientific creationists admit that their "predictions" cannot be materially traced from supernatural Creator to natural effect. Thus they must realize that their "predictions" are fiat assertions with a religious, not a scientific, base.

Scientific Creationism makes the same arguments that Morris used in his less widely known "Impact" article concerning the alleged circularity of biostratigraphic methods. In his textbook treatment of this issue, Morris tries to establish, again using selective quotations, three "facts" of science. Then, Morris attempts to discredit the methods which *are* used by biostratigraphers, again by the selective use of quotations that seem to contradict them. Morris's three "documented facts" are (1) that rocks are assigned geologic ages through fossils; (2) that fossils are arranged in order based on an assumption of evolution; and (3) that this fossil order, in turn, provides the main evidence for evolution (1974, pp. 95–96). From these "facts," Morris draws the conclusion that the evidence for evolution is based on the asumption of evolution.

Morris's three "documented facts" are three specious untruths, and the documentation consists of misrepresentations. Yet Morris's readers will undoubtedly be impressed by the sources from which he claims his support. For example, Morris cites Berry (1968), a well-known biostratigrapher, who apparently does believe that biostratigraphy is "based on the evolutionary development of organisms." In fact, the subtitle of his book *Growth of a Prehistoric Time Scale* is *Based on Organic Evolution.* How could Berry make such a claim? The only justification that Berry provides is that the "unique [evolutionary] events in time also provide markers or points in the general succession of fossils that may be obtained from the rocks and used as base-marks for time units" and that the "succession of faunas and floras seen in the rocks of the earth's crust is the product of . . . the evolution of organisms through natural selection" (1968, p. 42). All of this is perfectly true, but it is the mere material *existence* of the succession of fossils, not how it was formed, that is the basis of biostratigraphy and, together with lithostratigraphy, of the geologic time scale. Morris also quotes from Evernden et al. (1964) who stated that vertebrate paleontologists do sometimes, or did, rely on "stage of evolution" to determine the chronologic relationship of faunas. This statement is perfectly true, but it is also true that they did so in full knowledge of the theoretical difficulties such a method entails and that it is only a poor substitute for normal biostratigraphy. I will discuss this complex case in more detail later in this chapter.

Morris's third "fact" requires immediate treatment. Fossils do not provide the main evidence for evolution. This is a myth, not popularized, I hope, by paleontologists. The fossil record is only one of many examples of evidence for evolution. In fact, Darwin considered the fossil record to be evidence against his theory of evolution because of the lack of transitional fossils between species (although there are transitions between higher taxa). Darwin had to attribute the gaps in the record of fossil evolution to the poor quality of the stratigraphic record, a problem still with us today. He thought that evolution involved the gradual transition of one species into another, so he expected to find at least one example of such a gradual transition in the fossil record. Many modern paleontologists believe that evolution was not gradual, but erratic, and the model of "punctuated" evolution explains the fossil record's apparently negative evidence for Darwin's gradualism (see Godfrey, this volume). Thus Morris's view that fossils provide the main evidence for evolution is wrong, and his supporting quotation by Kerkut (1960,

p. 134) that "fossils . . . provided the main factual basis for the modern view of evolution" is equally erroneous. The fossil record was certainly not Darwin's guiding light, and paleontologists made almost no contribution to the modern view of evolution—the "synthetic theory"—although paleontologist George Gaylord Simpson argued that the fossil record was consistent with this theory. Some paleontologists today are suggesting that the fossil record is not consistent with the synthetic theory of evolution and are actively working to modify this theory with their own new data and hypotheses.

When Morris tries to discredit the methods that are used to date rocks, he misleads his audience about the most basic facts of biostratigraphy. He (1974) makes these points: (1) Strata are not dated by their adjacent strata; (2) Strata are not dated by vertical superposition; (3) Strata are not dated radiometrically. Contrary to Morris's assertion, rocks are very definitely dated by adjacent rocks, specifically by the principle of vertical superposition. That is, younger sediments were deposited on top of older sediments, so that when the sediments are lithified (turned into rock), the younger rocks will be above the older rocks. As discussed earlier, this principle is based on physical law, actualism, and corroborated empirical observation. It has never been shown to be false. But Morris points out that there are exceptions to the rule. "Vertical position," he asserts, "ought to provide at least a local relative chronology. The many cases of 'inverted order,' [sometimes in perfect conformity], however, make this rule apparently an unreliable guide" (1974, p. 133).

It is true that strata are sometimes found in "inverted order"—that is, that older rocks are sometimes found resting on younger rocks—but some earth movement must disturb the horizontal strata for this to happen. This can happen in two ways, both of which can be recognized by trained geologists: (1) a *thrust fault* thrusts older strata on top of younger; and (2) the strata are *overturned* (turned upside down) so that the older beds are above the younger. There are many ways to identify thrust faults and overturned strata in the field; in fact, there is a book that describes the features and structures that geologists use to determine the sequence of layered rocks without using fossils (Shrock 1948). Usually the order of the fossils is not used to recognize thrust faults or overturned strata, but occasionally biostratigraphy must be used for this purpose. However, no section whose rock sequence is ambiguous is ever used to "zone" fossils for biostratigraphic purposes. That is, biostratigraphers never depend upon rocks whose vertical sequence is geologically ambiguous to develop an ordinal time scale based on fossils. Furthermore, contrary to Morris's assertion, there are relatively few cases of "inverted order," and numerous cases of undisturbed correct order, so the principle of vertical superposition is a very reliable guide to the age of strata.

Finally, when Morris (1974, pp. 132–33) asserts that "perfect conformity" sometimes exists in "inverted" sequences, he is merely stating the obvious—that tilting, folding, and overturning will not disturb the conformity that existed in the strata prior to the disturbance. Older rocks are *never* found conformably resting upon younger rocks unless the rocks are overturned. To imply, as Morris does, that this diminishes the value of the principle of vertical superposition is to deceive the reader.

Morris also quotes Spieker:

> Rocks of any 'age' may rest vertically on top of those of any other 'age.' The very 'oldest' rocks may occur directly beneath those of any subsequent 'age.'
>
> Further, how many geologists have pondered the fact that lying on the crystalline basement are found from place to place not merely Cambrian, but rocks of all ages?
>
> [1956, p. 1805, as cited by Morris 1974, pp. 132–33]

Spieker's statement is perfectly true but irrelevant to Morris's argument. Erosion of rocks overlying the pre-Cambrian "crystalline basement" allows younger strata of all ages to be deposited directly on it. This is not an unusual occurrence. Gaps in the stratigraphic record are the rule, not the exception, but superposition, correlation, and one hundred seventy years of work by geologists have produced a highly refined geologic time scale that has been corroborated many, many times despite the gaps.

Morris's false claim that rocks are not dated radiometrically is treated in other chapters in this volume (see Abell, Brush). Radiometry is not the only means geologists have of dating rocks, but merely as an independent means of dating rocks it helps check ages derived from biostratigraphy; as discussed earlier, the agreement in ages from these two methods corroborates both.

Once Morris denigrates to his satisfaction the actual methods of geological dating, he is ready to deliver the familiar final blow. "How, then, are rocks actually dated?" he asks. "What is it that determines the geologic age to which a given rock formation is assigned? The answer is *index fossils!*" (1974, p. 134).

> In each sedimentary stratum certain fossils seem to be characteristically abundant: these fossils are known as index fossils. If in a strange formation an index fossil is found, it is easy to date that particular layer of rock and to correlate it with other exposures in distant regions containing the same species.
>
> [Ransom 1964, p. 43, as cited by Morris 1974, p. 134]

But how do index fossils provide geologists with specific geological ages?

> The answer to this question is *evolution!* That is, since evolution has taken place in the same direction all over the world, the stage of evolution attained by the organisms living in a given age should be an infallible criterion to identify sediments deposited in that age.
>
> [Morris 1974, p. 134]

Thus, according to Morris, rocks are dated by their index fossils, which can only be placed in a temporal sequence if evolutionary progression is assumed.

Index fossils do help determine the age of a given rock formation, but only by correlation from a type section of rock that is first defined to be a certain age. Fossils, by themselves, do not determine an age, so no attribute of theirs, including evolution, is necessary to date a rock. Creationists believe that geologists use a fossil's supposed "stage of evolution" to determine age, but nothing could be further from the truth. Although we observe a general tendency for life to evolve from simple to complex, this cannot be predicted for any single group of organisms, and the record shows simple and complex organisms living side-by-side in every geologic age. Rather than a trend or progression in a single direction, the fossil record shows life evolving in many different directions, only a few of which can be described, using relatively objective criteria, as simple to complex. It is therefore not inevitable that once life starts on a planet it will evolve into a complex, intelligent creature. It is thus rarely possible to use as nebulous a concept as "stage of evolution" to date rocks. Yet Morris seems to derive this principle from no less an authority than a former president of the Geological Society of America.

> That our present-day knowledge of the sequence of strata in the earth's crust is in major part due to the evidence supplied by fossils is a truism. Merely in their role as distinctive rock constituents, fossils have furnished, through their record of the evolution of life on this planet, an amazingly effective key to the relative positioning of strata in widely separated regions and from continent to continent.
> [Hedberg 1961, p. 499, as cited by Morris 1974, p. 134]

Morris takes special care to emphasize the quotation from Hedberg by repeating what he believes to be its key phrases. "How is the sequence of strata determined?" he asks (1974, p. 135). As Hedberg states, fossils are "an amazingly effective key to the relative positioning of strata"; this is what correlation is all about. But how is "such an amazing thing" done (1974, p. 135)? Morris repeats and italicizes the phrase "through their record of the evolution of life." But this is *not* the substantive phrase that describes the method. Hedberg said "through," not "by" their record. The phrase is only an irrelevant aside that Hedberg used rhetorically. The substantive phrase, which Morris ignores, is "merely in their role as distinctive rock constituents." This phrase correctly describes how fossils are used to perform correlation: fossils are material objects in rock strata whose presence alone, in unvarying sequence of succession, allows one to perform correlation. So the ultimate key to geologic dating is not evolution, as Morris wishes us to believe, but rather repeated fossil succession, an empirical fact. By the way, Morris (1973, p. 4), in using the above quotation to support his argument of circularity, omits Hedberg's essential substantive phrase without indicating its absence.

Morris also seems to derive his "stage of evolution" argument from Evernden and his colleagues (1964) and from Dunbar (1949). First he provides us with the following quotes:

> Vertebrate paleontologists have relied upon 'stage-of-evolution' as the criterion for determining the chronologic relationships of faunas. Be-

> fore establishment of physical dates, evolutionary progression was the best method for dating fossiliferous strata.
>
> [Evernden et al. 1964, p. 394, as cited by Morris 1974, p. 135]

> Fossils provide the only historical, documentary evidence that life has evolved from simpler to more and more complex forms.
>
> [Dunbar 1949, p. 52, as cited by Morris 1974, p. 135]

Then he draws the familiar conclusions that an evolutionary progression of life must be *assumed* in order to date strata and that sequences of strata derived in this fashion are then used to support evolution. That is, "the main evidence for evolution is the assumption of evolution!" (Morris 1974, p. 136).

It is true that Evernden and others believe that paleontologists use some knowledge about evolution—"stage-of-evolution" or "evolutionary progression"—to date fossiliferous strata. If one refers to Evernden's original paper, however, it becomes clear that Morris is misrepresenting Evernden by lifting quotes out of context. The context is vital in this case: mammal fossil sites in North America are characteristically widely scattered, stratigraphically ambiguous localities that usually resist the application of the principle of superposition to determine vertical stratigraphic relationships. Also, the fossil-bearing strata and adjacent strata rarely contain fossils for which an adequate biozonation exists since mammal fossils occur in nonmarine sedimentary rocks. Because of these two factors, it is very difficult to date the mammal-bearing beds and vertebrate faunas by conventional methods. European mammal-bearing strata, however, are frequently interbedded with marine strata that can be precisely dated. Because of this, a general mammal evolutionary sequence had been worked out, and it was the custom of North American vertebrate paleontologists to place their mammal fossils on this evolutionary sequence by using the criterion of stage-of-evolution of the various faunal elements that had excellent records, such as horses and rodents. They literally had no other choice. This procedure is obviously antithetical to standard biostratigraphy based on superposition, faunal succession, and correlation, and Evernden et al. (1964) were quite aware of the problems such a method entails. They state:

> The use of characterizing aggregates [of fossil mammals] is a simple application of "stage-of-evolution" or "biogenetic" correlation and

> suffers from all the errors and lack of refinement that may accompany such a discipline. [Although] correlations based upon "stage-of-evolution" of Cenozoic land mammals have proven to be refined and practicable . . . mammal fossils do not furnish a refined biostratigraphic zonation of strata. [p. 147]

The paper goes on to point out that this is because the

> samples frequently lack demonstration of superposition [and cannot therefore be] characterized or demonstrated stratigraphically . . . by standard criteria. [p. 153]

Vertebrate paleontologists were therefore making the best of a difficult situation by using evolutionary progression to date samples. They were fully aware that this method was anomalous when compared to normal biostratigraphy. The whole purpose of the paper by Evernden et al. was to test the mammalian chronology in North America by comparing it with the potassium-argon ages of volcanic tuffs and flows that occur within or near the mammal fossil-bearing strata. The ages provided by the North American mammalian chronology (which had been correlated to the European mammalian chronology by stage-of-evolution) and the radiometric chronology were determined by completely independent methods, so the degree of congruence of the two ages for each sample locality would test the precision of both methods. The authors found that "the potassium-argon dates . . . substantiate the geochronology employed by North American vertebrate paleontologists and stratigraphers. . . . In almost every case, they have given support to the proposed time-stratigraphic position of . . . mammal ages" (pp. 163–66)

This result actually corroborates the stage-of-evolution correlation method for at least one fossil group. In fact, this method would probably work *only* for Cenozoic fossil mammals because they evolved relatively rapidly during the Cenozoic, were abundant, and their evolutionary history is very well known to the extent that paleontologists could place any fauna in the evolutionary sequence with little ambiguity. This last factor discourages the stage-of-evolution method's use for other fossil groups. However, there are theoretical biosystematic objections to any claim that a group's "evolutionary history is known," so the method remains theoretically weak. A very few biozonations do use evolutionary criteria to perform correlation and age assignment, but these deal with fossil organisms whose stages of evolution or evolutionary progression can be redefined on the basis of superpositional material characteristics, thereby avoiding the theoretically invalid invocation of evolutionary criteria.

Morris misrepresents Dunbar by implying that Dunbar supports his claim that fossils provide the only evidence for evolution. First, this is not true; there are many sources of evidence for evolution. Second, Dunbar cannot be saying this, for he is speaking only about evolution from simpler to more complex forms, not about evolution in general. Third, Morris (1974, p. 135) misquotes Dunbar by omitting, without so indicating, the phrase "although the comparative study of living plants and animals may give very convincing circumstantial evidence" from the beginning of Dunbar's (1949, p. 52) quotation. Fourth, Dunbar is only moderately correct: cladistic analysis today provides evidence that is superior to fossil evidence for documenting evolutionary relationships. That is, the "comparative study of living plants and animals" by cladistic analysis (unknown to Dunbar in 1949) provides additional empirical evidence that life has evolved from simpler to more complex forms in the lineages where this phenomenon is manifested.

Morris's creationism is a powerful system of fallacious reasoning. Seemingly authoritative, but actually misappropriated, quotations from legitimate scientists and non-scientists are used to suggest compelling support for the creationist argument. To refute the argument requires no small effort. But the challenge is not one of powerful reasoning and evidence: it is one of deceit, half-truths, and distortions. I have taken it upon myself to refute one argument in detail to show the real nature of the creationist pretension to scientific knowledge. The creationists' method of discovering knowledge is the antithesis of the scientific method. They must resort to quotation out of context, illogical reasoning, and misrepresentation because the main evidence for creation is the assumption of creation.

History of the Origin of Biostratigraphy and the Geologic Time Scale

Whoever still doubts that fossils can be used to correlate and date strata *without any recourse to evolutionary assumptions or theory* should try this: assume for a moment that all fossil species really are specially created. We can now ask whether biostratigraphy would still be possible? The answer is simple: of course it would. The men who invented biostratigraphy and constructed the geologic column thought so, for they were all creationists! The great irony of the modern scientific creationist argument of circularity in the use of biostratigraphy and the geologic time scale is that the entire system was developed long before 1859 when Darwin published *The Origin of Species* and scientists became evolutionists. Men like Smith, Buckland, Murchison, Sedgewick, Lyell, Conybeare, Oppel, Cuvier, Brongniart, and D'Orbigny were all creationists, and many were catastrophists. These men, by their efforts in the first half of the nineteenth century, built the geologic time scale by correlating strata with fossils. These men were the authentic scientific creationists; they were also all deists or providentialists who believed in a First Cause, but thought that the province of science was to study the secondary causes in nature.

William Smith, the father of stratigraphy, was the discoverer of the principle of biotic succession, so we will briefly review the relevant points of his work (see Hancock 1977 for sources and references). The principles of initial horizontality and superposition had been known since the seventeenth century, although Smith had probably never heard of Steno (1669), their discoverer. By 1796 Smith had discovered that "it was a general law that the same strata were always found in the same order of superposition and contained the same peculiar fossils." His discovery spread in England in the first two decades of the new century and was used by others. Later, Smith published these important remarks:

> By the tables it will be seen which Fossils are peculiar to any Stratum, and which are repeated in others. . . . The organized Fossils which may be found, will enable him to identify the Strata of his own estate with those of others. . . . My observations on this and other branches of the subject are entirely original, and unencumbered with theories, for I have none to support. [1817, pp. iv–vi]

This last sentence is the ultimate refutation of the modern creationist claim that biostratigraphy assumes evolutionary theory: Smith denies it in his own words! Hancock notes that in Smith's method, "the 'Strata' had first been delineated by mapping, and then their characteristic fossils collected" (1977, p. 4). This observation is the key to the method of biostratigraphy, which is, as I have previously emphasized, based entirely on lithostratigraphy: the empirical sequence of stratigraphic succession.

Conclusion

The creationists often claim that if man believes he is descended from an ape, he will act like it. I believe that I am descended from an apelike creature, but between that ancestor and myself lies an immense journey that nature reveals to us only by our great effort. It is a journey of physical and cultural evolution, of conflict and cooperation, of moral instinct and moral learning, of growth of self-consciousness and awareness of self-responsibility, of fear and wonder at an uncaring universe and the prospect of death, of a oneness with nature and alienation from nature, and of a struggle to know the truth. Man's evolutionary journey has prepared him to face life and the universe with acceptance in the face of meaninglessness and hope in the face of ignorance. There is, indeed, grandeur in this view of life. Thomas Henry Huxley stated that he would rather have an ape for an ancestor than a man who would use his restless intellect to obscure scientific questions by aimless rhetoric, eloquent digressions, and skilled appeals to religious prejudice. Apparently, some things never change, and antievolutionists continue to use their restless intellects to formulate factual distortions and specious arguments to defend their creationist doctrine, for their knowledge of the "truth" justifies such behavior. Huxley did not inquire into the ancestry of his opponent, but today the arrogance and self-righteousness of the true believers can be explained. They regard themselves as being created in the image of God, and act like it.

REFERENCES CITED

Ager, D. V. 1973. *The nature of the stratigraphic record.* New York:, Wiley.
Berry, William B. N. 1968. *Growth of a prehistoric time scale, based on organic evolution.* San Francisco: W. H. Freeman.

Dunbar, Carl O. 1949. *Historical geology.* New York: Wiley.
Evernden, J. F., Savage, D. E., Curtis, G. H., and James, G. T. 1964. Potassium-argon dates and the Cenozoic mammalian chronology of North America. *American Journal of Science* 262:145–98.
Gish, Duane, T. 1973. *Evolution? The fossils say no!* San Diego: Creation-Life Pubs.
Gould, Stephen J. 1981. Evolution as fact and theory. *Discover* (May), pp. 34–37.
Hancock, J. M. 1977. The historic development of concepts of biostratigraphic correlation. In *Concepts and methods of biostratigraphy,* eds. Erle G. Kauffman and Joseph E. Hazel, pp. 3–22. Stroudsburg, Pa.: Dowden, Hutchinson & Ross.
Hedberg, H. D. 1961. The stratigraphic panorama. *Bulletin of the Geological Society of America* 72:499–518.
Kerkut, G. A. 1960. *Implications of evolution.* Oxford, Eng.: Pergamon Press.
Kitts, D. G. 1974. Paleontology and evolutionary theory. *Evolution* 28:458–72.
Morris, Henry M. 1973. Geology and the flood. ICR Impact Series, no. 6.
—. 1974. Ed., *Scientific creationism.* San Diego: Creation-Life Pubs.
—. 1977. Circular reasoning in evolutionary geology. ICR Impact Series, no. 48.
Nelson, Gareth J. 1969. Origin and diversification of Teleostean fishes. *Annals of the New York Academy of Sciences* 167:18–30.
O'Rourke, J. E. 1976. Pragmatism versus materialism in stratigraphy. *American Journal of Science* 276:47–55.
Parker, Gary E. 1980. *Creation: the facts of life.* San Diego: Creation-Life Pubs.
Ransom, Jay E. 1964. *Fossils in America.* New York: Harper & Row.
Schaeffer, B., Hecht, M. K., and Eldredge, N. 1972. Phylogeny and paleontology. *Evolutionary Biology* 6:31–46.
Schindewolf, Otto H. 1960. Stratigraphische Method and Terminologie. *Geologische Rundschau* 49:1–35.
Schrock, R. R. 1948. *Sequence in layered rocks.* New York: McGraw-Hill.
Smith, William. 1817. *Stratigraphical system of organized fossils.* London: E. Williams.
Spieker, E. M. 1956. Mountain-building chronology and the nature of the geological time-scale. *American Association of Petroleum Geologists Bulletin,* vol. 40, no. 8, pp. 1805.
Steno, Nicolaus. 1669. *De solido intra solidum naturaliter contento dissertationis prodromus.* Florence: Printing shop under the Sign of the Star.
Valentine, James W. and Campbell, Cathryn A. 1975. Genetic regulation and the fossil record. *American Scientist,* vol. 63, no. 6, pp. 673.
West, R. R. 1968. Paleontology and uniformitarianism. *Compass* 45:216.
Wilder-Smith, A. E. 1980. *The natural sciences know nothing of evolution.* San Diego: Master Books, Creation-Life Publs.
Wysong, R. L. 1976. *The creation-evolution controversy.* Midland, Mich.: Inquiry Press.

13

Humans in Time and Space

By C. LORING BRACE

Prehuman Fossil Primates

To the creationist, the questions of human origins and development loom especially large because of their acceptance of the biblical view that human beings are the object and end of creation and because of their commitment to the literal truth of the account of creation in Genesis. That there are mutually incompatible statements concerning the creation right at the beginning of Genesis (Alter 1981) does not seem to have bothered them, but the evidence supporting an evolutionary interpretation of human beginnings has been called into question even though no creationist has ever bothered to study the fossil material available. Nor have the creationists made any effort to discover what is known by even a superficial perusal of the existing and easily obtainable primary source publications. Despite this, promoters of the creationist view, in a book prepared for public school use, declare "our purpose here is to show there is no evidence supporting the assumed evolutionary descent of man from an apelike ancestor" (Morris 1974, p. 177). This declared "purpose" has nothing whatever to do with the scientific treatment of the available data but simply reflects their prior commitment to the defense of sectarian religious dogma (Morris 1963, p. 77; 1975, p. 183).

A quick review of the primate and hominid fossil evidence and interpretations will show that the pat dismissal by the creationists does less than justice to what has become a burgeoning mass of data, all of which fit comfortably within an evolutionary framework and are hard to account for in any other way.

Human beings, anthropoid apes, monkeys, tarsiers, lemurs, and lorises are grouped in the Linnean order Primates (Clark 1950; Simons 1972). Initially this classification was based solely on the recognition of shared anatomical similarities. Linnaeus, as a good creationist, interpreted the similarities as evidence of God's plan. Over a century later, after Darwin taught scientists to look at the world in the perspective of time and adaptive change, it was realized that the similarities among members of a classificatory unit reflected community of descent (Simpson 1953, 1961).

The spectrum of living primates runs from that most modified and aberrant species, *Homo sapiens,* to prosimian forms that are so little different from non-primate insectivores that scientists have been arguing for a century about their correct classification (Luckett 1980). The important thing, in reality, is not the "correct" pigeonhole but the fact that they represent a condition intermediate between the two orders and suggest to us the kind of evolutionary change by which primates could have diverged from the generalized mammalian stem.

The fossil record confirms this since the earliest primates to appear are prosimians only (Clark 1960; Simons 1972). The gradual change through time and identifiable diversification has been shown with admirable clarity by Gingerich whose plot of molar dimensions is reproduced in figure 1 (1976, p. 16). Whether the rate of change through time follows the fits-and-starts expectations of the proposers of the "punctuated equilibria" model of evolutionary change (Eldredge and Gould 1972; Gould and Eldredge 1977) or the path of "evolutionary gradualism" (Gingerich 1976) or a combination of the two (Brace 1981), the demonstrable steps show how microevolutionary changes accumulate to produce macroevolutionary results.

If the Paleocene (up to 65 million years ago) and Eocene (up to 55 million years ago) primates appear to be prosimians in their proportions

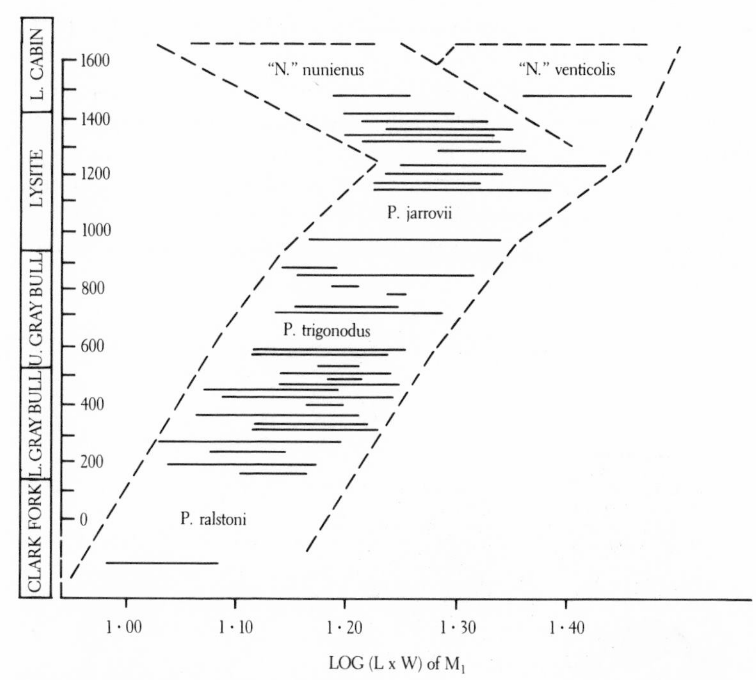

Figure 1. EVOLUTION OF THE PRIMATE GENUS *NOTHARCTUS* FROM *PELYCODUS* SEEN IN A SIXTEEN-HUNDRED-FOOT-THICK SEQUENCE OF EARLY EOCENE BEDS IN THE BIG HORN BASIN IN WYOMING
The horizontal extent of each line is based on the cross-sectional area of the lower first molar of specimens found at a given level. Some 255 individuals are represented. [Drawn after Gingerich 1976, figure 7., with permission.]

of snout to braincase and their dental adaptation to an insectivorous diet, creatures that we would regard as belonging to the monkey grade of organization appear in the subsequent Oligocene epoch (23 to 38 million years ago). The best documented of these, *Aegyptopithecus xeuxis,* was an arboreal quadruped about the size of a smallish dog. It had a tail, grasping hands and feet, and a brain that was somewhat enlarged when compared with those of the prosimians (Fleagle, Simons and Conroy 1975; Radinsky 1973; Simons 1967). Molar tooth crowns, unlike the earlier Paleocene and Eocene primates, were low and rounded like those of recent fruit-eating primates. All of this is completely monkeylike (Fig. 2), but when one looks at the patterns of cusp arrangement on the molar teeth, they are quite different from those of modern monkeys but absolutely indistinguishable from those of modern anthropoid apes—and human beings.

Although *Aegyptopithecus* has been regarded as a "dental ape," it had the body plan of a monkey. Instead of having the chest, shoulders, and arms modified for hanging—suspensory locomotion—it was an orthodox

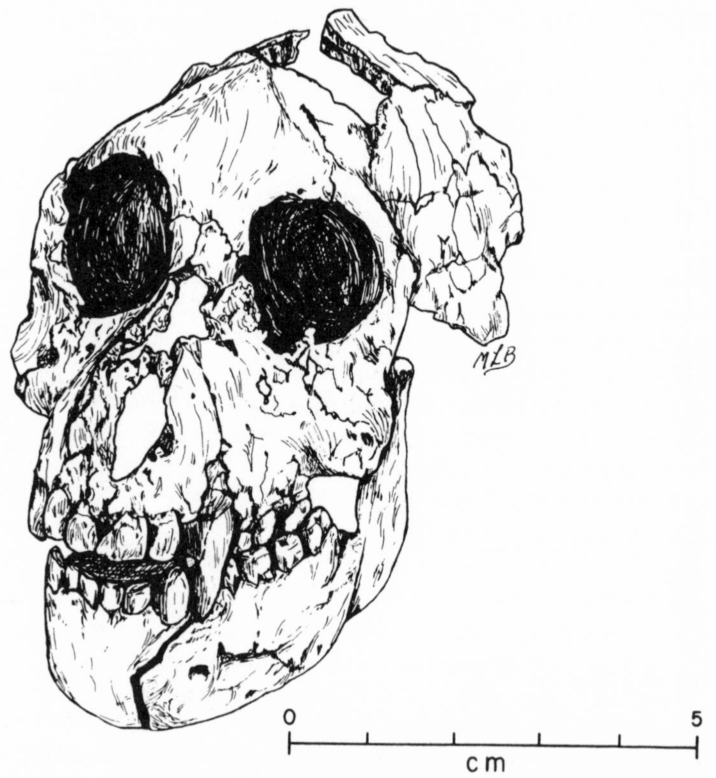

Figure 2. *AEGYPTOPITHECUS XEUXIS,* A FOSSIL PRIMATE FROM THE UPPER OLIGOCENE OF THE FAYUM BASIN IN EGYPT
This represents an evolutionary advance over the earlier Prosimians, but it comes before either true apes or primitive humans and is the best example of a form that could have given rise to both. [From Brace et al. 1979, p. 15.]

arboreal quadruped. All told, it provides a splendid representation of the ancestral condition from which modern apes—and humans—descended (Brace and Montagu 1977; Wolpoff 1980).

So far, the remains of *Aegyptopithecus* are known only from deposits in the Fayum depression west of the Nile in Egypt near the northern edge of the African continent. In the following Miocene epoch (from 5 to more than 20 million years ago), however, skeletal remains from obviously related creatures are known from France, Greece, Hungary, Turkey, Pakistan, China, and Africa (Pilbeam 1980; Wolpoff 1980*b*). All of these resemble the first one discovered in France in 1856, *Dryopithecus fontani*, so they have been categorized as Dryopithecines. There is evident regional and ecological differentiation, and some of what used to be taken as specific variation appears to be simply an expression of male/female size difference (Greenfield 1972, 1977).

Most of our evidence is from jaws and teeth, of which there are now a substantial and increasing number, but there are some scraps of faces and skulls, and recent finds of limb bone fragments suggest that some had become locomotor as well as dental apes. Others remained quadrupeds like the earlier *Aegyptopithecus*. Somewhere within this Dryopithecine complex is the probable ancestor of later true hominids. Debate continues, for example, on whether specimens called *Ramapithecus* from the Siwalik hills in the northern part of the Indian subcontinent are a proper genus or whether they should be included within the range of variation of the previously named *Sivapithecus* (Pilbeam 1968; Greenfield 1972, 1977). Whatever designation finally achieves accepted recognition, most anthropologists feel comfortable with the idea that the specimens that constitute that group make good representatives of the form that was ancestral to both more recent apes and to human beings.

Australopithecines

Dryopithecine fossils are spottily distributed over a 10-million-year time span from about 18 to about 8 million years ago. Then there is another gap in the fossil record, and, until recently, there were no known hominid fossils older than 2 million years. At 2 million years, we find fossils from a series of East and South African sites that allow us to trace the development of one of the more remarkable transformations visible in paleontology. It involves no less than the emergence of human form

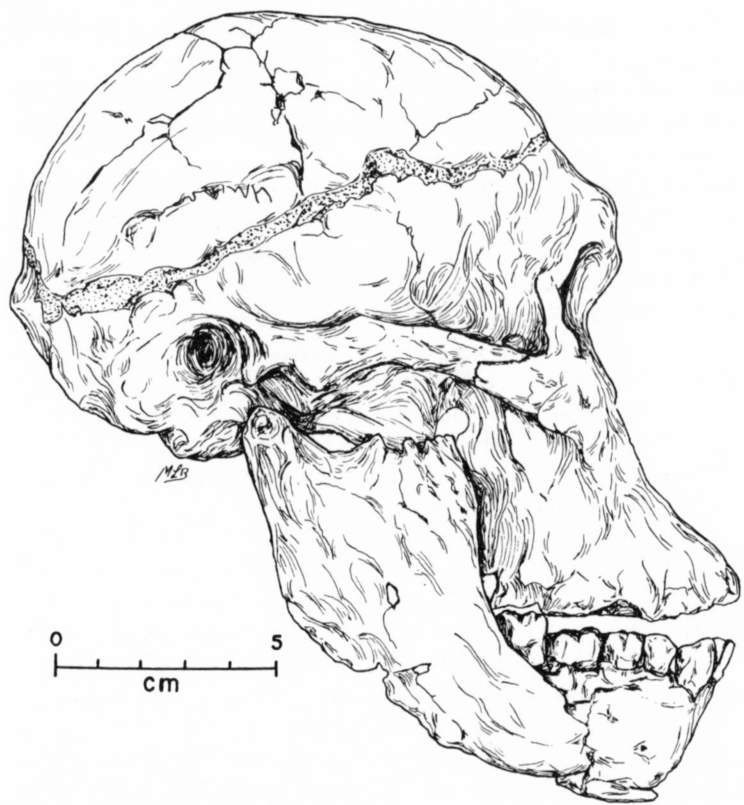

Figure 3. AN AUSTRALOPITHECINE SKULL AND JAW FROM STERKFONTEIN IN THE TRANSVAAL REGION OF SOUTH AFRICA
Although these come from different individuals, they show the features that were characteristic of *Australopithecus africanus*, the probable human ancestor of 2 million years ago. [From Brace et al. 1979, p. 37.]

out of what are very obviously nonhuman antecedents. These fossils have been included in the genus *Australopithecus*.

The first representative of this group to be discovered was an immature specimen found in South Africa in 1924. It consisted of a complete face, jaws, and deciduous teeth, and an endocast of the braincase from which the skull bones had broken away and been lost. Later on, further specimens from other sites in South Africa confirmed the impressions conveyed by the original find.

Brain size was approximately the same as that found in the large modern anthropoid apes. The crowns of the molar teeth had the same cusp and fissure pattern visible in the Miocene Dryopithecines and also in modern apes and humans. The canines, however, did not project in a functional way beyond the occlusal level of the rest of the teeth. In this trait, the Australopithecines resembled human beings and differed from all other primates. Finally, the spinal chord entered the skull at the bottom and not toward the back as in quadrupedal mammals, and this, confirmed later by discoveries of pelvic and lower limb bones, showed that the Australopithecines stood erect. In fact, careful anatomical and biomechanical studies have demonstrated that they were erect-walking bipeds whose mode of locomotion was not different from that of modern *Homo sapiens* (Lovejoy, Heiple and Burstein 1973). This is the cautious, checked, and tested conclusion of all of the qualified scholars who have worked with the available original material (Jenkins and Fleagle 1975; Lovejoy 1974, 1975, 1978; Tuttle 1967, 1969).

The direct evidence, then, gives us a picture of *Australopithecus* as a terrestrial biped with an ape-sized brain and possessing ape-sized teeth that, however, occlude in a completely human fashion. The locomotor adaptation is so different from that of a typical pongid—ape—that the creature is placed in the same family, the Hominidae, with modern human beings, and referred to informally as a hominid. However, this does not make it a human being. With a brain that is only one-third the capacity of the modern human average and teeth that are double the bulk, the Australopithecines warrant separate generic designation—*Australopithecus.* The implications of the anatomical, ecological, and archaeological evidence have all been weighed and considered with care (Brace 1979*a*, 1979*d;* Isaac and McCown 1976; Jolly 1978; Tuttle 1975; Wolpoff 1980*b*), and despite the differences of opinion among those who have studied the original material, no one doubts the fact that the Australopithecines present that "mosaic of advanced and ancestral characters" that "always" characterizes an evolutionary intermediate (Mayr 1971, p. 50). Within the spectrum of the Australopithecines, then, we find a picture of an intermediate between the pongid and the hominid condition that is just as convincing as that provided by *Archaeopteryx* of an intermediate between the reptilian and the avian condition.

Contrast this with the cavalier treatment accorded by creationists, based only on hearsay evidence, secondary sources, and without any firsthand familiarity with the original specimens—"these creatures were nothing but apes" (Gish 1974, p. 16). The "scientific" creationist account, referring to the same unreliable sources, concludes with unwarranted conviction that *Australopithecus* was a "long-armed, short-legged, knuckle-walker." Finally, "*Australopithecus* not only had a brain like an ape, but he also looked like an ape and walked like an ape. He, the same as *Ramapithecus*, is no doubt simply an extinct ape" (Morris 1974, p. 173).

So far, all of this discussion deals only with discoveries that had been made prior to the 1970s. Subsequent work by Dr. Donald C. Johanson and his associates in Ethiopia and by Mary D. Leakey and colleagues at Laetoli just south of Olduvai Gorge in Tanzania have extended our knowledge of the Australopithecines back another million and a half years to a time 3.5 million years ago (Johanson, White and Coppens 1978; Johanson and White 1979; M. D. Leakey et al. 1976). The oldest and the most extensive of this material is that discovered as a result of Johanson's efforts at the Hadar site in the badlands of the Afar depression in north central Ethiopia. There he discovered the relatively complete skeleton of a small adult female Australopithecine. The tale of her discovery and interpretation is engagingly told for the reading public in the book called *Lucy* after the informal name given to this, his most famous, find (Johanson and Edey 1981). Further field work uncovered more than a dozen additional specimens. These have been studied in exemplary comparative detail in Johanson's laboratory at the Cleveland Museum of Natural History where they were available for the inspection of interested qualified scholars for several years before they were returned to the Ethiopian government for permanent storage in Addis Ababa. The results are being put in print as expeditiously as possible, and enough is now available for us to be able to assess the nature and significance of these, the oldest hominids of which we have substantial evidence (Johanson, White and Coppens, 1982; Ward, Johanson, Coppens, 1982; White and Johanson 1982).

All who have examined Johanson's extensive collection and the roughly contemporary but scrappier material from Mary Leakey's work at Laetoli (White 1977, 1980) agree that these are Australopithecines, that is, members of genus *Australopithecus*. All also agree that in a series of traits these early Australopithecines are more primitive than those that are dated to approximately 2 million years ago. The legs are relatively short and the arms are relatively long, although the anatomy of the pelvis, femur, tibia, and ankle and foot bones leaves absolutely no doubt that they were erect-walking bipeds. There is nothing to indicate that they could have been knuckle-walkers. Brain size was right in the range of variation of anthropoid apes and possibly, on the average, slightly smaller than in later Australopithecines. The position of entry of the spinal chord into the skull is in-between that characteristic of humans and anthropoid apes.

The dentition also shows enough pongid features so that if one simply had the better part of the skull, face, and teeth, an expert could quite reasonably conclude that the creature could not be distinguished from

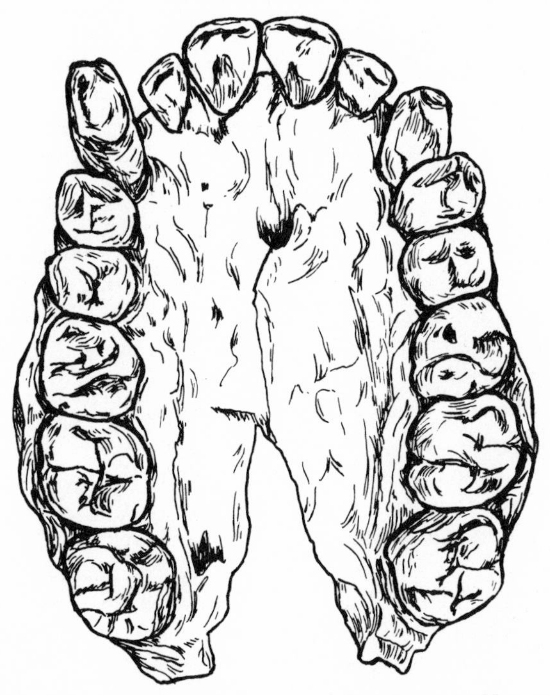

Figure 4. AN EARLY AUSTRALOPITHECINE PALATE, AL 200, FROM HADAR IN THE AFAR DEPRESSION OF ETHIOPIA
This is perfectly intermediate between the apelike and the human conditions. It is the bipedal body plan of these creatures that ranks them as hominids and not as pongids. [Drawn after Johanson and White 1979.]

a fossil ape. The sides of the dental arch are relatively straight and parallel rather than being oriented as part of a parabolic arch as is typical for human beings. The incisors are broad, and there is a noticeable gap —a diastema—separating them from the canines in the fashion visible in anthropoid apes but not in humans. The canines themselves have crowns that are more robust than those of any subsequent hominid. In their unworn form, they project slightly beyond the occlusal surface of the rest of the teeth, although they wear flat just as human canines do. Finally, the lower first premolars, just behind the canines, show a mesio-buccal to disto-lingual (outer-front to inner-back) elongation and ridge that is very close to the form of the sectorial (cutting) premolar typically (but not universally) found in nonhuman primates and quite different from the bicuspid crown form typical for humans. In pongids, this elongate ridge slices upward against the inner surface of the projecting upper canine—the two forming the arms of a scissorlike shearing mechanism. However, an investigation with a scanning electron microscope of the wear on the early Australopithecine teeth has shown that the wear pattern is not produced by the typical pongid shearing motion but more by a flat crushing usage (Ryan 1980).

This primitive configuration of pongid and hominid traits has led the discoverers and describers of these early Australopithecines to assign them to a new species. If the first Australopithecine to be discovered is properly *Australopithecus africanus,* the early ones, they suggest, should be *Australopithecus afarensis.* Not all scholars agree. I have to confess that, although I have had the opportunity to handle both the Ethiopian and the South African material with which it is being compared, and although I agree with virtually all of what its describers say in regard to its tendency to be more primitive in a series of traits, I am not convinced that the differences are pronounced enough to warrant separate specific recognition. My reasoning goes like this: The Neanderthal skeletal remains of fifty thousand and more years ago all differ from modern *Homo sapiens* in that they are more primitive in an equal number of traits, but few students regard these differences as being of enough importance to warrant distinction at more than the subspecific level.

Our disagreement is merely a matter of the assignment of names. This is based on the judgment of individual scholars and is a trivial matter, but it does point up an issue of fundamental significance. In an evolutionary continuum, change occurs more or less gradually through time.

At the early and late ends of such change, everyone agrees that different names are justified, but when one form slowly transforms into another without break, the point where the change of name is to be applied is a completely arbitrary matter imposed by the namers for their convenience only—it is not something compelled by the data.

Australopithecus into Homo

If the foregoing is true as we watch the early Australopithecines become the late Australopithecines, the same is equally true for the transformation of *Australopithecus* into *Homo.* It is a fascinating story with an interesting twist or two, and, although we can see the major outlines, we need more information before we can be sure of the what and the when.

As has been noted above, *Australopithecus africanus* was present in South Africa 2 million years ago, and it was only slightly modified from the *Australopithecus* of 1.5 million years earlier. In the next half million years, however, things changed rather dramatically. At 1.5 million years ago, we have evidence for the presence of two hominids in the fossil record. Both are erect-walking bipeds, and, as far as we can tell from what is preserved, there was essentially no difference between them from the neck on down. The head is a different story. One has the Australopithecine ape-sized brain but displays a substantial size increase in the teeth, jaws, and attachments for the chewing muscles. The other has a braincase that is nearly double that of the average Australopithecine and molars that are only half the size of its large-toothed compatriot.

The large-toothed small-brained biped is called *Australopithecus boisei* after the type specimen discovered by Mary and Louis Leakey in Olduvai Gorge in 1959. Jaws, teeth, and at least one more complete skull have since been found elsewhere in East Africa with one particularly complete specimen being found by the Leakey's son Richard at Ileret east of Lake Turkana in northern Kenya (see fig. 5). The relatively small front teeth, the greatly expanded crushing surfaces of the premolars and molars, and the pronounced development of the attachments for chewing muscles suggest that *A. boisei* focused its subsistence activities on the tough seeds, nuts, and roots of the East African savannah terrain that, judged by the nature of the deposits where it is found, had been its habitat. *A. boisei,* then, is regarded as having been a specialized

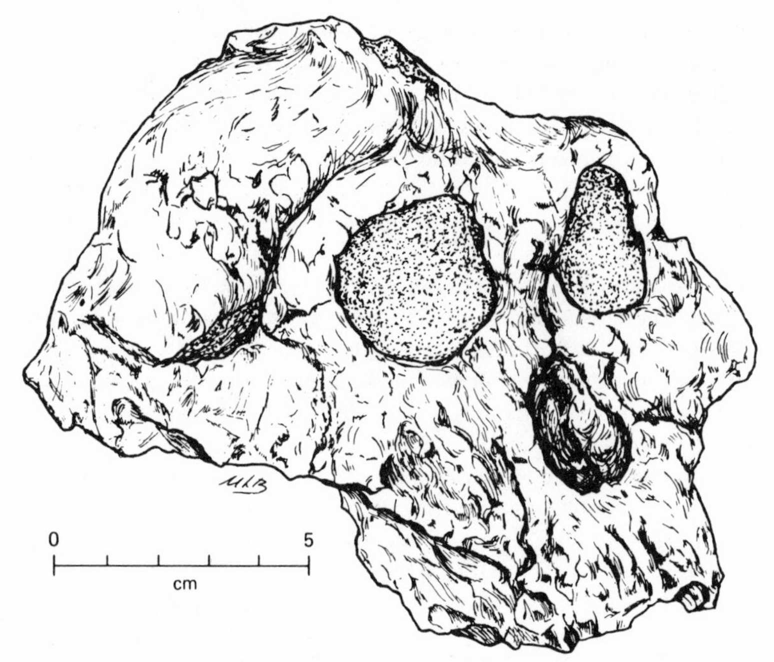

Figure 5. A ROBUST LATE AUSTRALOPITHECINE, ER 406, FROM ILERET EAST OF LAKE TURKANA IN NORTHERN KENYA
[From Brace et al. 1979, p. 44.]

vegetarian that diverged from the ancestral *A. africanus* condition after 2 million years ago and flourished for a time, disappearing for good around 1 million years ago.

Its contemporary at 1.5 million years ago also appears to have diverged from the ancestral *A. africanus* condition after 2 million years ago and to have then taken quite a different tack. Brain size had expanded to the extent that it exhibited a range of variation that overlapped the lower limits of what is found in modern *Homo sapiens.* At the same time, tooth size had reduced so that it overlapped the upper end of the *Homo sapiens* range. This same configuration characterized the hominid fossils found in Java in the 1890s by the Dutch physician Eugene Dubois, and at Zhou Kou Dian (Choukoutien) near Beijing (Peking) in China by W. C. Pei and his colleagues in the late 1920s and 1930s. These fossils are all accepted as belonging in our own genus, *Homo*, with the specific designation, given by Dubois to his 1891 and 1892 finds, *erectus*.

Suggestive fragments of skulls, jaws, teeth, and a few limb bones had been found in contexts that have been a little hard to date in both East and South Africa ever since the 1950s, but finally, in 1975, fieldwork under the direction of Richard Leakey east of Lake Turkana in Kenya uncovered a specimen that was assigned the number ER 3733 (fig. 6). This could be dated to at least 1.3 million years ago from potassium-argon determinations in an overlying volcanic tuff. Brain size has been shown to be close to nine hundred cubic centimeters, which overlaps the modern range where normal individuals exhibit cranial capacities from near eight hundred to over two thousand cubic centimeters. Here at last

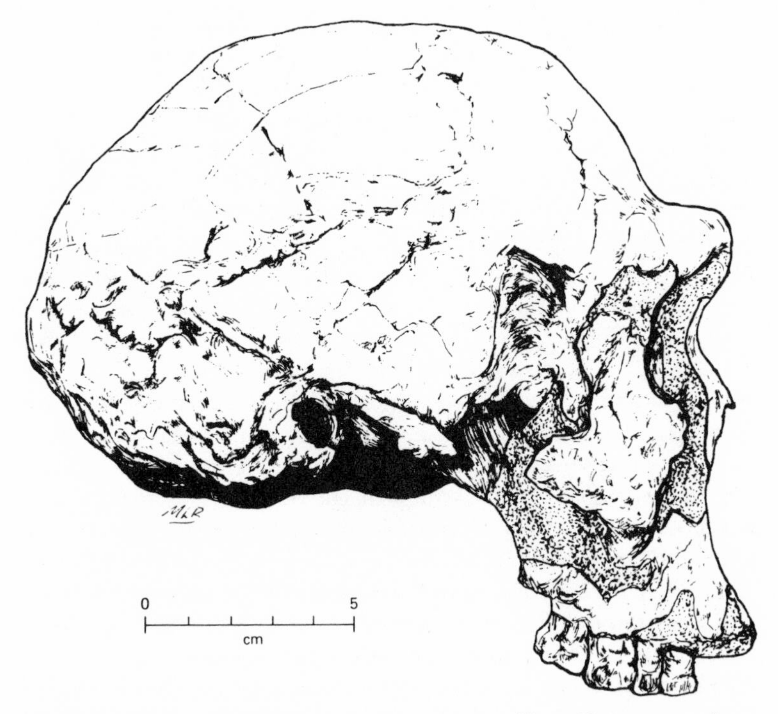

Figure 6. AN EARLY *HOMO ERECTUS*, ER 3733, FROM KOOBI FORA EAST OF LAKE TURKANA IN NORTHERN KENYA
[From Brace et al. 1979, p. 66.]

was conclusive evidence that *Homo erectus* had been a contemporary of the late robust Australopithecines. Some have taken this to mean that *Australopithecus* could not have been the ancestor of *Homo,* but, as a consideration of the available evidence will show, there is no reason why the 2-million-year-old *A. africanus* could not have given rise to both *A. boisei* on the one hand and *Homo erectus* on the other.

In East Africa, at Olduvai Gorge in Tanzania and east of Lake Turkana in Kenya, the field work conducted by the Leakey family has led to the discovery of hominid fossils in that time span between 2 and 1.5 million years ago. These display just that mosaic of primitive and advanced features that Ernst Mayr has predicted should occur in evolutionary intermediates (1971, p. 50). Not unexpectedly this has given rise to a cloud of terminological confusion. The first such specimen to be discovered was by the Leakeys, early in the 1960s, in Bed I at Olduvai Gorge at a level dated to more than 1.75 million years old. This find, numbered OH 7, included a lower jaw in which the second molars (twelve-year molars in a modern human) had just erupted and two partial and flattened skull bones. This specimen was compared to others from as much as half a million years later in Bed II of Olduvai Gorge, and the assemblage was given the label of *Homo habilis* in a burst of somewhat premature enthusiasm (L. S. B. Leakey, Tobias and, Napier 1964). No definitive description of this specimen has yet been published, nor has the required careful morphological and quantitative analysis yet been done, and it would appear that the assignation of the new species name was improperly proposed (Brace, Mahler, and Rosen 1973).

The jaw and its teeth are completely indistinguishable from a typical specimen of *A. africanus.* The crushed adolescent cranial bones, however, according to several considered efforts at reconstruction, indicate a brain size that was either at the small end of the *H. erectus* (Holloway 1980) or the large end of the Australopithecine range of variation (Wolpoff, 1981).

Then, in 1972, at the site of Koobi Fora in the East Turkana region of northern Kenya, Richard Leakey's group discovered a skull and facial skeleton, given the number ER 1470, that has been the center of conflicting claims and controversy ever since (fig. 7). Its discoverers hailed it as proof that "true *Homo*" existed nearly 3 million years ago (R. E. F. Leakey 1973; Leakey and Lewin 1977) and therefore as proof that *Australopithecus* and possibly even *H. erectus* could not have been

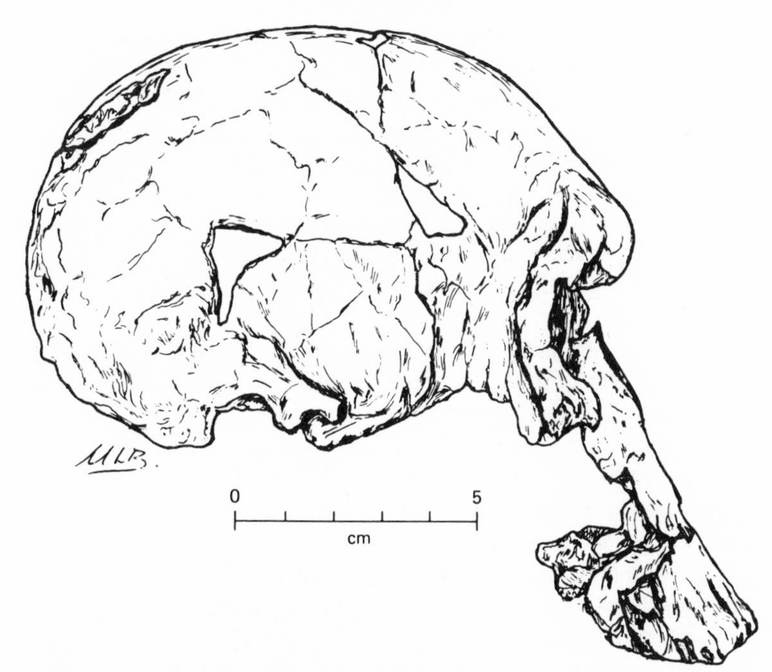

Figure 7. ER 1470, AN INTERMEDIATE BETWEEN *AUSTRALOPITHECUS* AND *HOMO* FOUND IN STRATA OF ABOUT 1.8 MILLION YEARS AGO IN THE KOOBI FORA REGION EAST OF LAKE TURKANA, KENYA
[From Brace et al. 1979, p. 54.]

our ancestors. This was a theme that Louis Leakey had been pursuing since early in the 1930s, long before Olduvai Gorge had begun to yield its fossil treasures (cf. Brace 1979*c*, 1981; Johanson and Edey 1981), and it has been hailed with satisfaction by the creationist whose preconceptions lead them to rejoice in anything that would confound the picture of an orderly course of human evolution (Gish 1974, p. 21, 1979; Morris 1974, p. 176).

As it happens, there are flaws in the claims concerning both the date and the anatomical significance of the specimen. Elements in the fauna at the Koobi Fora site are present at other well-dated African sites at an age of between 1.6 and 1.8 million years (Cooke 1976; White and Harris 1977), and when tuff specimens from Koobi Fora were reanalyzed with the application of proper laboratory procedures, it became clear that the original figure for ER 1470 was incorrect; its true age was also between 1.6 and 1.8 million years. (Curtis 1975; Curtis et al. 1975). In addition, the "true *Homo*" status of the skull also was claimed with somewhat more enthusiasm than a careful appraisal of the evidence would support. As it happens, the long, somewhat "dish-shaped" face and the enormous molar root sockets look perfectly Australopithecine. The brain case itself also looks Australopithecine in form, but, at 750+ cc., it is exactly in-between *Australopithecus* and *Homo* (Holloway 1975).

It is taking nothing away from the significance of ER 1470 to say that it does not support the contentions of its finders. It *does* fit between two major categories—the creationists would say "kinds"—of hominid, and it provides a gratifying picture of one of the steps by which *Australopithecus* became transformed into *Homo*. Because it shows such a mixture of features, it is not surprising that the authorities who have studied it have been unable to agree on a named pigeonhole for its assignment. According to one respected authority, it is best called *Australopithecus habilis* (Walker 1980). Another recognized that it really belongs to the Australopithecine grade of organization, but nevertheless assigned it to genus *Homo* (Wolpoff 1980*b*, p. 165). Nor is ER 1470 alone in its intermediate status. Other specimens such as ER 1813 and OH 16 also show evidence on the one hand of a reduction in tooth size and on the other of an expansion of brain size that make positive categorical assignment an uncertain thing. All of this, however, is just what one would expect at the time when one genus was in the course of developing into another.

The Cultural Record

There is another major category of evidence that appears and becomes of increasing importance during the time when *Australopithecus* gives rise to *Homo*. This is represented by the appearance of stone tools, the imperishable parts of a cultural record that extends without break from 2 million years ago right up to the present day.

At first, these stone tools are a pretty unimpressive lot—pebbles the size of a human fist or smaller with a flake or two removed. These were called Oldowan by Louis Leakey who first recognized them in Olduvai

Gorge in Tanzania in the 1930s. We would not even know they were the product of some deliberate selective agent if it were not for the fact that they appear in deposits at such a distance—and even upstream—of the nearest possible rock sources that no environmental agency could have transported them there. They occur in the deposits at Olduvai Gorge going right back to the earliest strata at 2 million years.

Given a terrestrial biped lacking enlarged canine teeth, the existence of some kind of hand-held defensive weapon would hardly be a surprise. However, the fossil record shows that our biped has an antiquity of at least 3.5 million years, and, given the locomotor development already achieved by that date, there is reason to suspect that its bipedal adaptation must predate that by at least another half million years. Yet stone tools do not appear until 2 million years ago. One can guess, in the absence of any evidence, that the actual defensive weapons utilized since the first development of bipedalism were fabricated from perishable materials—the proverbial pointed stick, for example. When the first stone tools appear at about 2 million years—that is, right down at the bottom of the dated strata in Olduvai Gorge—we can suspect that they were being used for something other than simple defense.

Of course we cannot *know* with certainty what they were all being used for, but their frequent association with the broken bones of various kinds of mammals at Olduvai and elsewhere suggests that they were butchering tools wielded by their users to get through the hides and joints of the mammals with whose remains they occur. This is most dramatically suggested at the HA site at the Koobi Fora locality east of Lake Turkana where a quantity of such tools is found in association with the partially butchered skeleton of an extinct hippopotamus all silted over in the deposits of an ancient river delta (Isaac 1976).

During the half-million-year time represented by the accumulation of Bed I and the lower part of Bed II at Olduvai Gorge, the tools show an increasing sophistication and diversity, and the nature of the animal bones with which they are associated suggests that the users had progressed from the butchering of creatures that had died of natural causes to the deliberate killing of increasingly large animals for food (Isaac 1976). What we see evidently is the record of the development of a tool-making animal from a gatherer and occasional scavenger to a deliberate big game hunter. The changes in cranio-facial anatomy that occur as *Australopithecus* becomes *Homo* can be comfortably accounted for

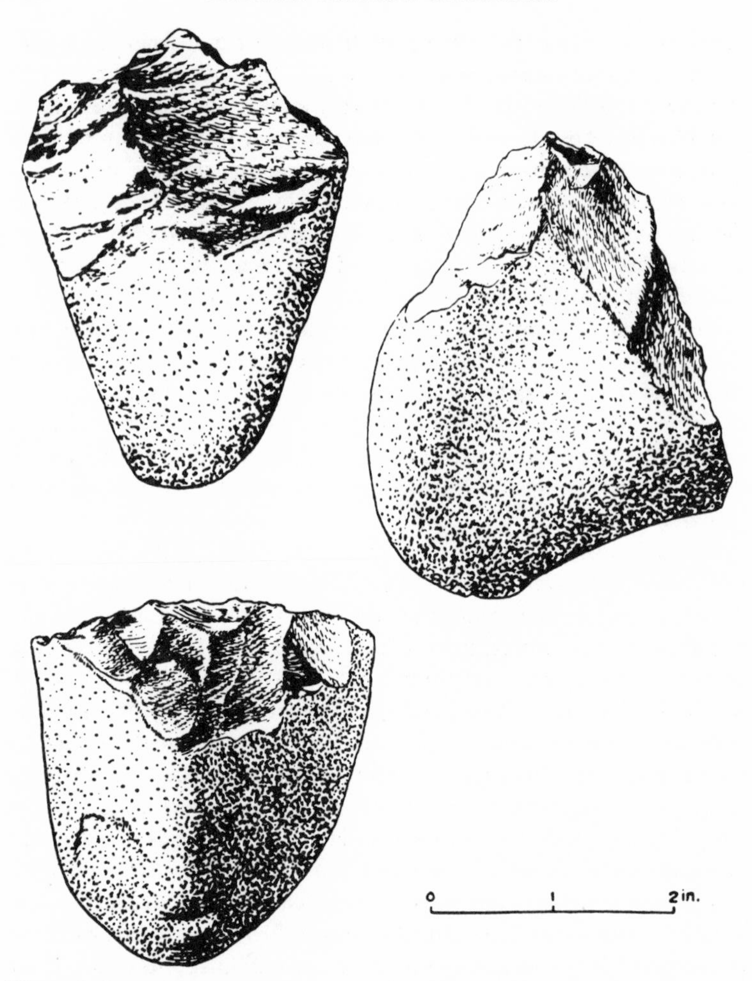

Figure 8. OLDOWAN TOOLS FROM THE LOWER PLEISTOCENE, BED I, OF OLDUVAI GORGE IN NORTHERN TANZANIA
[From Brace et al. 1979, p. 26.]

in the perspective of the changes in the forces of natural selection that accompany such a transition in subsistence strategy (Brace 1979*a*, 1979 *c*).

It is just such circumstances that would lead to a marked increase in brain size. Certainly the other aspects of hominid anatomy by themselves are pretty unpromising for a would-be predator. As modes of locomotion go, bipedalism is not notably rapid, and a bipedal primate could hardly expect to capture much of anything by outrunning it. Nor could a creature whose canines did not project beyond the level of its incisors pose much of a threat by trying to bite its would-be prey if it did succeed in catching up with it. Its only hope would be in learning so much of the seasonal, species, and individual peculiarities of the various members of the animal world that it could plan their capture from ambush with the aid of hand-held tools and weapons. Such are the circumstances under which we would predict a relatively rapid increase in the size of the organ by which such planning is accomplished, namely the brain, and it surely is no accident that the available skeletal finds show that hominid brain size effectively doubled during just that period when big game hunting became an important hominid subsistence activity.

Once stone tools appear in the prehistoric strata 2 million years ago, they continue to provide us with an unbroken record of the activities of our ancestors. The chances of fossilization and preservation of any given individual of a creature who lived at such a low population density as the early hominids are pretty small, and it is not surprising that hominid fossils are such rare and spotty phenomena. Stone tools, however, are made of imperishable materials, and once they become a regular part of the hominid cultural repertoire, the archaeologists who study them can trace those activities to which they pertain and tell us what regions of the world their makers were occupying and what changes they made in their way of life through time.

The occasional glimpses we get of the fossilized fragments of the makers of that unbroken cultural tradition enable us to check on the course of human evolution with the conviction that the earlier fossil hominids are indeed the ancestors of the ones who come later and show those modifications that foretell the emergence of modern form. The unbroken continuity of stone tools from the levels at the bottom of Bed I in Olduvai Gorge to the time five to six thousand years ago when written accounts begin to provide us with an articulate picture of human doings serves as an abundant check and confirmation of the anatomical continuity provided by the interrupted sequence visible in the course of the fossil record.

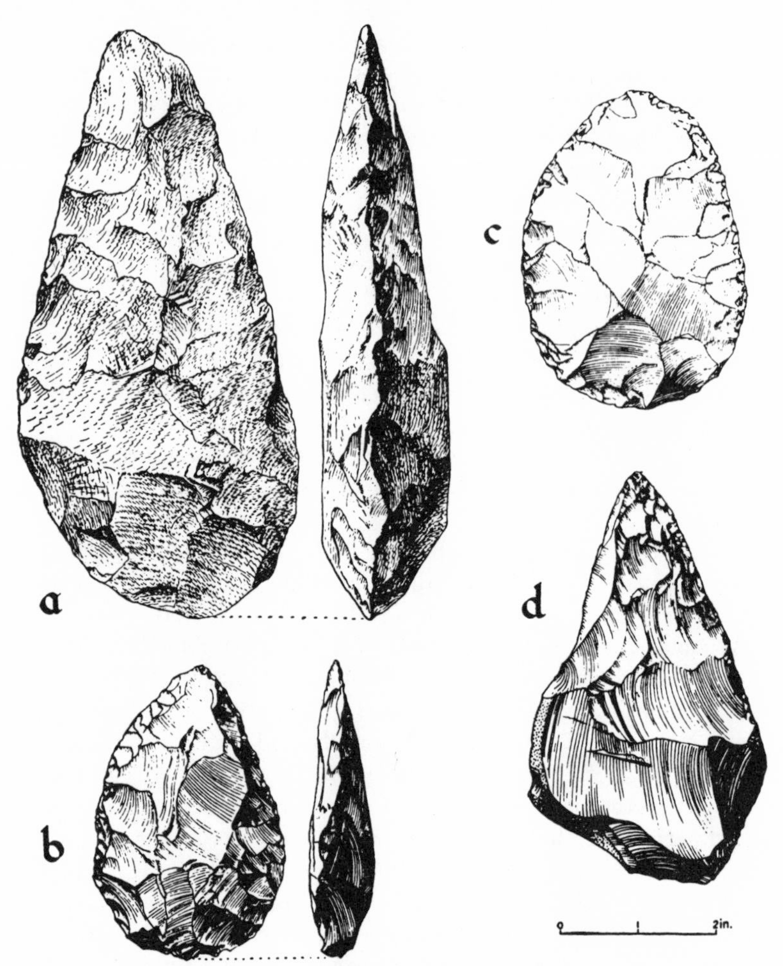

Figure 9. ACHEULIAN TOOLS FROM MIDDLE PLEISTOCENE DEPOSITS OF (A) KENYA, (B) FRANCE, (C) ISRAEL, AND (D) ENGLAND
[From Brace et al. 1979, p. 58.]

The evidence from archaeology shows that a basic and successful hunting and gathering mode of subsistence had been established before the beginning of the Middle Pleistocene three-quarters of a million years ago. This remained relatively unchanged until just before the onset of

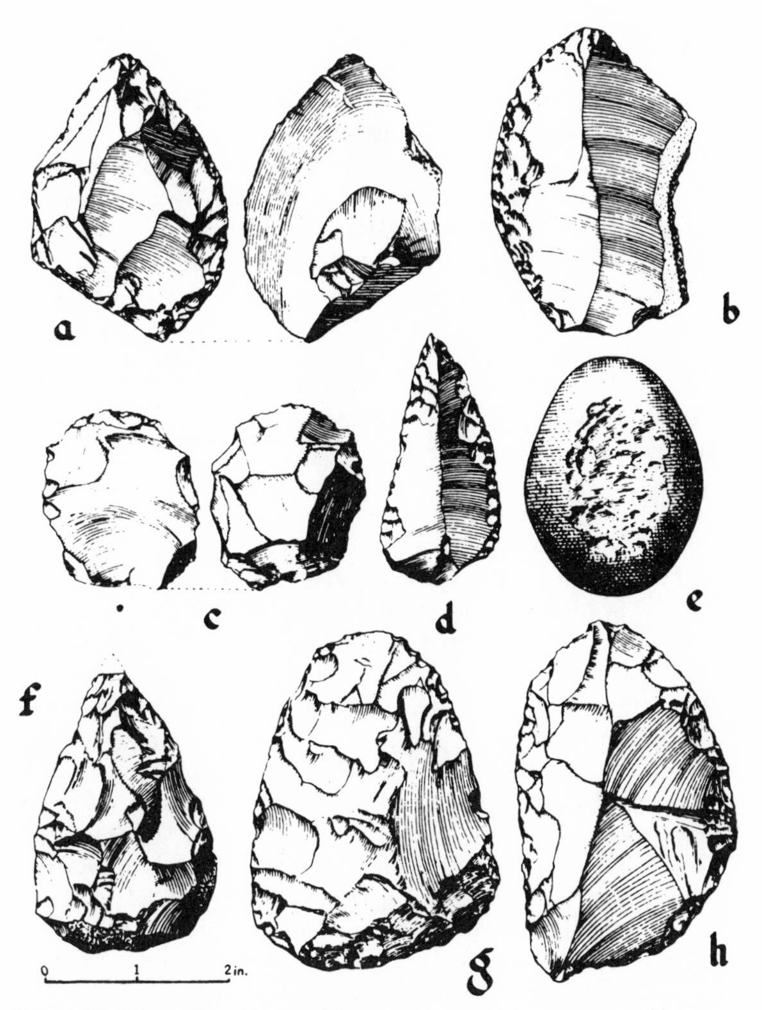

Figure 10. Upper Pleistocene Mousterian tools used by the Neanderthal hunters of Europe during the last glacial stage about fifty thousand years ago
[From Brace et al. 1979, p. 102.]

the last glacial episode about 100,000 years ago. Subsequently, technological refinements occurred at an ever-increasing rate of speed. Just before the end of the last glacial period, in several different parts of the world, there is evidence—the appearance of grinding stones, mortars and pestles, and other tools—that people had discovered how to prepare grains in such a way that these previously indigestible parts of the plant kingdom could be used as food. This provided human populations with a vast new subsistence resource, and, shortly, with the development of techniques of plant tending—i.e., agriculture—the vast increase in human numbers began that has led ultimately to what we now perceive as the "population problem."

One of the more indefensible predictions offered by the recently resurgent creationists movement is the idea that the origin of "civilization" is contemporaneous with the origin of "man" (Morris 1974, p. 13). Thus they have predicted that "man's agriculture and other basic technologies are essentially as old as man himself" (Morris 1975, p. 152). Such a position can only be maintained by deliberately choosing to ignore more than a century of archaeological research.

The 1.5 to 2 million year extent of the hunting and gathering way of life has been repeatedly tested and confirmed (Lee and DeVore 1968; Butzer et al. 1975). The gradual development of agriculture after the end of the Pleistocene ten thousand years ago has also repeatedly been tested and confirmed (Reed 1978). The discovery of various metals in sequence—copper, bronze, iron—is also well attested to not only by archaeological work but also in the accounts of written history as they begin to accumulate. This the creationists simply reject out of hand and declare, in the absence of any consideration of the actual data and their interpretation, that these archaeological and historical data "correlate perfectly" with the "Biblical model" (Morris 1975, p. 109). Once again, their prior commitment to a faith, which they justify on the grounds of sectarian religion, means that the evidence of science is rejected without a serious hearing.

Erectus to Sapiens

The evidence for the distribution and activities of *Homo erectus* increases in abundance after the beginning of the Middle Pleistocene some seven-hundred-thousand years ago. If the earliest reliable evidence

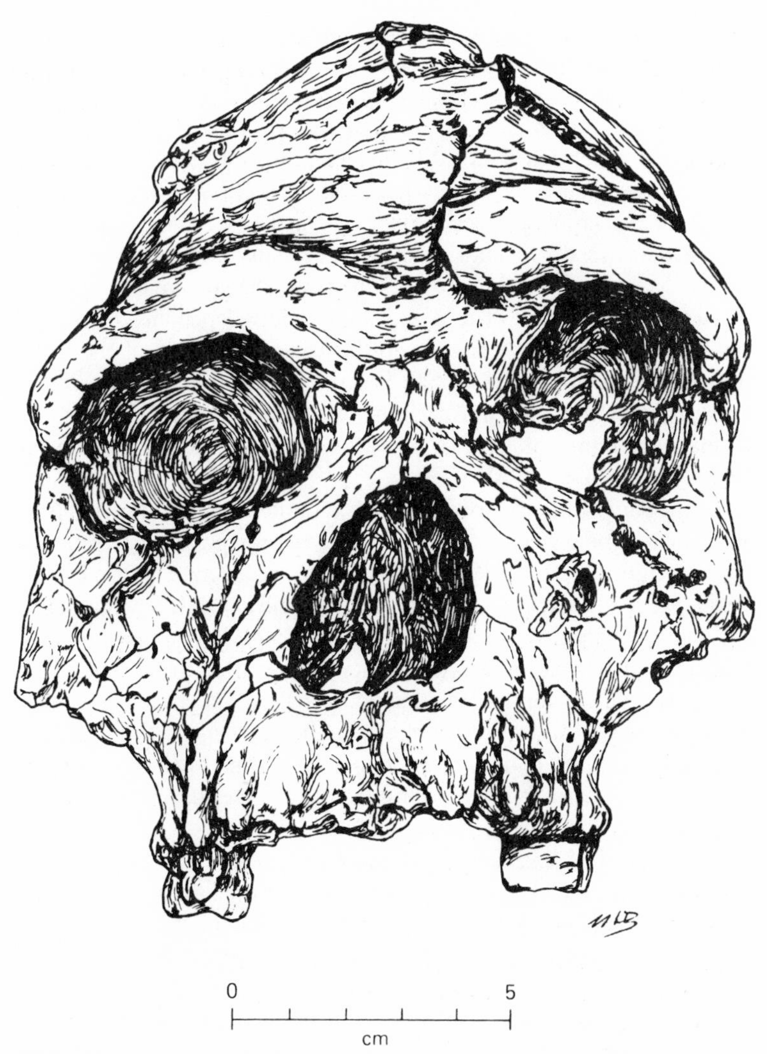

Figure 11. A skull of *Homo erectus,* Arago XXI, from Middle Pleistocene deposits in southern France
[From Brace et al. 1979, p. 79.]

we have is from East Africa in the Lower Pleistocene, it is clear that subsequently *H. erectus* spread eastward across the tropics of the Old World and was inhabiting Java by the beginning of the Middle Pleistocene. No specimens have yet been found in India, but there is an abundant archaeological record there that bears testimony to the hominid presence (Fairservis 1975; Khatri 1963; Sankalia 1978).

One of the Javanese specimens was the first to be found, and it was this that was used to identify the species *erectus* as distinct from and earlier than sapiens (Dubois 1894). Curiously, the creationists' reaction to this specimen and what it indicates differs depending on which creationists one reads. According to the associate director of the Institute for Creation Research in San Diego, both the original Java *erectus* discovery and the later similar forms found near Beijing (Peking) in China were representatives of a category that "must . . . have been simply a giant ape" (Gish 1974, p. 17; 1979, p. 123).

At the same time, the director of the same institute was willing to grant that "*Homo erectus* was a true man, but somewhat degenerate in size and culture" (Morris 1974, p. 174). This "degenerate" status is then attributed to possible "inbreeding, poor diet and a hostile environment" without any consideration of the evidence at all. Furthermore, a misreading of an account in a scientific news column led the author-editor to the completely unfounded conclusion that *Homo erectus* had been alive and well in Australia only ten thousand years ago. Since "modern man had already been in existence long before," the conclusion is drawn that ancestral status is therefore ruled out, and *erectus* "is more likely a decadent descendant" (Morris 1974, p. 144).

Now even if *H. erectus* had survived, isolated in Australia until ten thousand years ago, there is no reason at all why the known early *erectus* populations in China, Java, Africa, and Europe could not have given rise to the *sapiens* populations that follow them in time in just those areas. In this case, however, the news report merely indicated that the skulls in question only "suggest that . . . some *Homo erectus* genes lingered in Australia" (Morris 1974, p. 174, quoting from *Scientific American* Oct. 1972, p. 48). To make the jump from "some *Homo erectus* genes" to "these *Homo erectus* skulls" is an irresponsible misuse of what is at best a secondary source.

The material in question was excavated from the edge of Kow Swamp in the middle reaches of the Murray River drainage basin between South Australia and New South Wales (Thorne and Macumber 1972). The jaws and teeth preserved a Middle Pleistocene degree of robustness apparently as a result of the selective forces that were associated with a way of life that exploited the late survival of giant marsupials in southern Australia, but a fully *sapiens* expansion of the brain was also clearly present (Brace 1980; Thorne and Wolpoff 1981). The Kow Swamp people, then, were unequivocally *Homo sapiens.* Nor was there anything even faintly "decadent" or "degenerate" in their exuberantly robust skeletal form.

This does bring up the matter of the adaptive shift by which *erectus* was converted into *sapiens,* which was the expansion of the average size of the braincase from approximately one thousand to near fifteen hundred cubic centimeters. There is a standard deviation of about one hundred cubic centimeters, so there is a considerable overlap of the normal modern and *erectus* ranges of variation. This means that our decision concerning the designation of given specimens will be a statistical one that will misassign a considerable number of individuals. One could predict then that there should be a considerable degree of disagreement among professional anthropologists in regard to the designation of a number of specimens in the time span where the final transformation was taking place.

The evidence shows that average brain size had reached modern levels just before the onset of the last glaciation about 100,000 years ago (Holloway 1981*a,* 1981*b*). From that point on, brain size remains the same and all the hominids we know are included as members of *Homo sapiens.* The earlier of these still preserve Middle Pleistocene levels of skeletal robustness, which include strongly developed jaws and teeth and marked muscle attachments. Because of the markedly differing male and female subsistence roles, there is reason to suspect that there was a greater degree of sexual dimorphism than is true for more recent human populations (Brace 1973; Brace and Ryan 1980). As a result, one might expect that late *erectus* females might be assessed as *sapiens* by some appraisals.

It is matters such as these that have led to the confusion in specific labels attached to specimens predating the last glacial episode, like the Swanscombe and Steinheim skulls in Europe—both probably female

(Howells 1980; Wolpoff 1980*a*). Then there is the example of the collection of skulls found on the banks of the Solo River at Ngandong in Java early in the 1930s. Their first describer originally regarded them as a Far Eastern equivalent to the evolutionary grade represented by the European Neanderthals (Weidenreich 1940). He had recognized their cladistic affinity to the earlier *erectus* specimens in the Far East, and he continued to utilize both *Homo* and the older generic designation of *Pithecanthropus* (Weidenreich 1947). This assessment has been reiterated when the specimens were referred to as *Homo erectus* in the most recent review of the material, although the basis was an appraisal of details of cranial muscle attachments and bony reinforcements without any particular concern for the appraisal of adaptive grade (Santa Luca 1977). The fact of the matter is that the Solo individuals lived right at the time when the *erectus-sapiens* transition was taking place, and it is not the least bit surprising that anthropologists have shown so much indecision and disagreement in regard to their precise placement.

The hominid fossil record provides as good a picture of both major and minor evolutionary transitions as we possess. We still do not have a picture of how a quadruped was transformed into a biped. That would require the discovery of fossil material in the time span between 4 and 8 million years ago, a gap that is still unfilled. Between 3.5 and 2 million years ago, all observers agree that the only hominid to be seen is properly regarded as *Australopithecus,* a terrestrial biped with an ape-sized brain. Whether the earliest form is specifically distinct from the later one is a matter of opinion, but, in any case, neither would be recognized as having attained fully human status.

By 1.5 million years ago, genus *Homo* had made its appearance characterized by twice the brain size of *Australopithecus.* Between 2 and 1.5 million years ago, however, there are a number of hominid fossils that show all sorts of combinations of partial braincase expansion and dental reduction. Clearly we see pieces of evidence of a major evolutionary transformation under way—in fact, we see the development of human from nonhuman form. Over the next million plus years, the gradual expansion of the braincase by another 50 percent accomplished the transformation of *Homo erectus* into *Homo sapiens.* The change was gradual enough so that it is virtually impossible to make a hard-and-fast designation for many of the specimens that occur towards the end of the *erectus* and the beginning of the *sapiens* span.

Modern Human Evolution

One often hears it said by anthropologists and nonspecialists alike that, once modern form had been achieved, evolution stopped—"we" had arrived, so to speak—and the only further changes were in the cultural sphere. This viewpoint shows a certain kinship with the creationist assumption that "if evolution is taking place today, it operates too slowly to be measureable, and, therefore, is outside the realm of empirical science" (Morris 1974, p. 5).

Presumably, scientific observers cannot live long enough to watch true evolution taking place, or they cannot project themselves back in time to observe it. But a check of the fossilized remains of the past *can* give scientists the needed data, and, as we have already noted in the case of hominid fossils, the information supporting an evolutionary picture of human emergence is as good as or better than that for any species of which we have record. Furthermore, there is evidence to show that gradual changes have occurred during the time since the beginning of written records, scarcely more than four thousand years ago.

For example, the faces, jaws, and teeth of the early farmers of Jericho, whose walls tumbled later at the blast of Joshua's trumpet, were slightly larger and more robust than those of the modern inhabitants of the same general area (Brace 1977). The same kind of difference distinguishes the first English farmers from the average citizens of London in the seventeenth century (Brace 1979*b*). The differences are not dramatic and the magnitude amounts to less than a 5 percent average change, which is not by itself statistically significant, but the fact that the direction of difference is the same in all the traits measured is in itself highly important.

Biologists refer to changes of such a nature as "microevolution." Although the time through which microevolutionary changes are observed is not sufficient to produce the transformation in "kind," or "macroevolution," that creationists refuse to credit to the evolutionary process, all one needs is more time. As has been written, "macroevolution is nothing but microevolution over longer time spans" (Alexander 1978, p. 101). And despite the contorted denials of the creationists, time is available in abundance.

The early agriculturalists of the Neolithic, four, five, and six thousand years ago, and more, are preceded by people classified as Mesolithic.

This was the state in cultural development when the techniques of food procurement and plant utilization were developed that made possible the subsequent development of farming in the Neolithic. While Mesolithic people are technically classified as "hunter-gatherers" like all preceding members of genus *Homo* during the Pleistocene, the sophistication that they applied to their subsistence activities was an order of magnitude greater than that of earlier times. Face, jaw, and tooth size during the Mesolithic was approximately 10 percent more robust than that of their evident modern descendants (Brace 1979*b*).

If we then look back another ten or fifteen thousand years to the Upper Paleolithic precursors of the Mesolithic, face, jaw, and tooth size is more robust still, being between 20 and 30 percent more robust than the modern condition in Europe and Asia. We still regard these in-

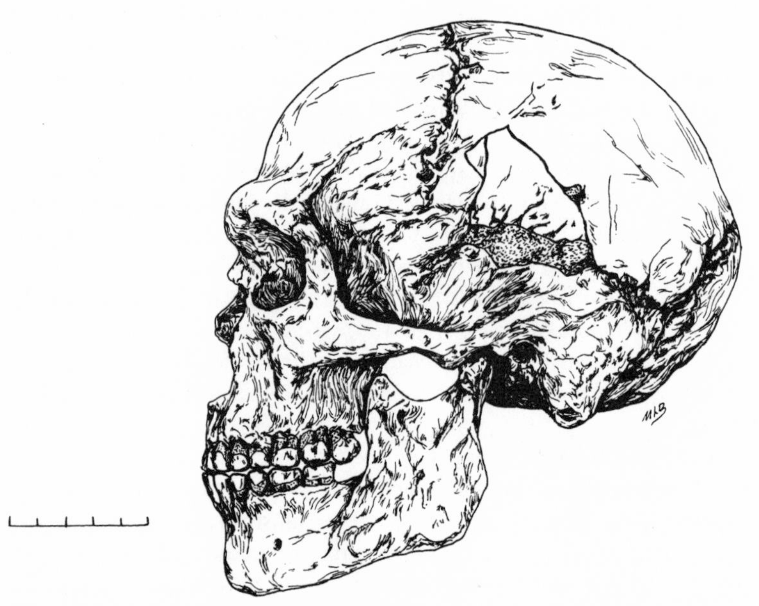

Figure 12. AN UPPER PALEOLITHIC EUROPEAN, PREDMOST III FROM CZECHOSLOVAKIA ABOUT THIRTY THOUSAND YEARS AGO
Brow ridges, jaws, and teeth are substantially more robust than in modern Europeans but less so than in the Neanderthals of fifty thousand years ago. [Drawn from a cast by Mary L. Brace.]

dividuals as being "modern" in form, although the adjective "primitive" is sometimes attached to them. However, another ten thousand years earlier, and another 10 percent more robustness of tooth, jaw, and face, and we perceive enough difference from the modern condition so that we use a separate name for that configuration—Neanderthal—and many anthropologists have balked at accepting this as representing the modern human ancestor (Boule and Vallois 1957; Howells 1974, 1975).

These are the obvious microevolutionary changes that can be seen in human dento-facial form during the last 100,000 years. Projecting them back 2 million years would easily predict a configuration that would differ from modern form to such an extent that we would recognize it as generically different from *Homo*—i.e., a difference in the creationists' "kind"—even if we had not yet found the fossils to prove it. But we *have* such fossils, and they are called *Australopithecus*. The overall change clearly is a case of macroevolution, and we have the evidence to show how it was accomplished by long-continuing microevolution.

When one then looks at the differences in dento-facial robustness in the various populations of modern *Homo sapiens*, one can observe a range of variation that is more than just the expression of chance individual differences. In aboriginal Australia, for example, fully Middle Pleistocene levels of tooth size continue right up to the present (Brace 1980), while maximum degrees of reduction can be found in such places as South China, the Middle East, and the pre-European people in the Valley of Mexico (Brace 1978, 1979*b*, 1980; Brace and Mahler 1971). The maximum amounts of reduction all occur in just those areas where the archaeological evidence shows that food-processing techniques—from earth ovens to seed grinders to pottery—have been employed for the longest period of time. The dento-facial changes, then, represent the effects of alterations in the intensity of the forces of selection.

Features of the human skeleton can be checked from the present right back into the past, but this is impossible to do for those aspects of human form that cannot be fossilized. Skin color, blood types, characteristics of digestive physiology, and a roster of other traits wherein people vary cannot be checked for early historic or prehistoric populations. We can, however, employ the "natural experiment" approach so successfully utilized by Charles Darwin, and we can add to this the information derived from modern laboratory and hospital observations to help us understand the meaning and probable course and place of development

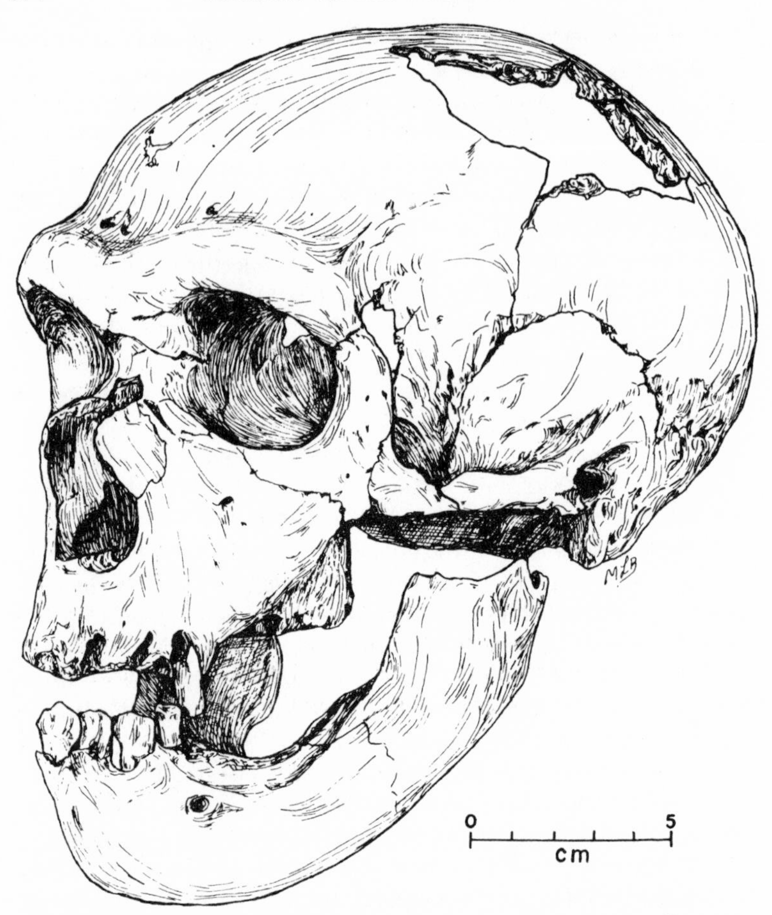

Figure 13. A EUROPEAN NEANDERTHAL FROM LA CHAPELLE-AUX-SAINTS IN SOUTHERN FRANCE ABOUT FIFTY THOUSAND YEARS AGO
Full modern brain size has been achieved, so this is formally recognized as *Homo sapiens,* but Middle Pleistocene levels of robustness are still apparent. [From Brace et al. 1979, p. 117.]

of an increasing number of traits that differ between one population and another.

One of the best understood of these is the hemoglobin form that causes sickle-cell anemia. Clinically, this is recognized in America as a phenomenon that causes misery and early death, principally in people of African ancestry. What is less well known is the fact that that particular form of hemoglobin enables its possessors to cope with the effects of an especially noxious form of malaria (Livingstone 1973).

Although many in the medical world have assumed that there is something uniquely African about sickle-cell anemia and that the presence of the contributing gene is a marker of African ancestry, the situation is quite otherwise. What controls the amount of the gene responsible for sickle-cell anemia in a population is the intensity of malaria infestation and how long it has been in effect in the area under consideration (Livingstone 1958).

As it happens, the introduction of agriculture to sub-Saharan Africa over the last two thousand years or so produced conditions that favored the propagation of the mosquito that transmits the malarial parasite at the same time that they promoted an increase in the size of the human population to the point where it could sustain a continuous level of parasite infestation. There are good reasons to suspect that both falciparum malaria and the hemoglobin that allowed people to cope with its impact, however unpleasant the side effects, were introduced into Africa from east of the Mediterranean. The selective force represented by the mortality caused by the malaria parasite led to an increase in the frequency of the gene that helped people survive malarial infection. This also had the less desirable consequence of increasing the number of those who suffered from the miseries of sickle-cell anemia.

By examining the figures for mortality caused on the one hand by malaria in those without the adaptive hemoglobin and on the other hand by sufferers of sickle-cell anemia, it is possible to predict just how long it took that balance of selective forces to produce the percentages of normal and sickle-cell producing hemoglobin in the various places studied. These time figures correspond nicely to those that can be independently checked by the evidence from linguistics and history (Livingstone 1958). When all the evidence is assessed, we have as good an example as we can get of microevolution having produced what we now perceive as significant differences in modern human populations over a matter of only a few thousand years.

There are some other traits also that are associated with variations in the intensity of identifiable selective forces. Peripheral circulation is maintained in arctic populations where frostbite is a threat (Fisher 1961;

So 1980; Steegmann 1975); higher aldosterone levels help the body retain salt despite sweating in tropical populations (Gleibermann 1973; Macfarlane 1973); long-time residents in the drier and/or colder parts of the world have elongated nasal passages to moisten inspired air (Weiner 1954; Wolpoff 1968); and human skin pigment is concentrated in those populations who live in parts of the world where the sunlight exposes them to the greatest amounts of ultraviolet radiation (Grosch and Hopwood 1979; Urbach 1969). These and other traits show degrees of variation in different populations proportional to the length of time that they have inhabited regions where the relevant selective forces have existed with varying amounts of impact.

Conclusion

Recently, the proponents of a view that they label "scientific creationism" have argued that "the creation-cataclysm model of earth history fits all the known facts of man's history much better than the evolution model and it recognizes that man's agriculture and other basic technologies are essentially as old as man himself" (Morris 1975, p. 152; Morris 1974, pp. 171–201). Such a view can only be maintained, however, by either ignoring or denying virtually all of the data and their implications accumulated by biologists, paleontologists, archaeologists, and anthropologists as a result of a century and more of increasingly carefully checked and substantiated work.

The creationists have complained that it is unfair for a teacher to present only the scientific evidence for evolution. To do this, they claim, is "a process of indoctrination, and the school degenerates into a hatchery of parrots" (Morris 1974, p. 178). But the scientific evidence for evolution can be examined, questioned, and tested—as the creationists themselves have been doing for over a century in the effort to discredit it—which is a splendid demonstration of just how science works. Creation, on the other hand, as its own supporters freely grant, is "inaccessible to the scientific method." No "scientific experiment" can be devised to test it since "the Creator does not create at the whim of a scientist" (Morris 1974, p. 5). It is creation and not evolution, then, that is "indoctrination," and if students are required to spend equal time learning it in the public schools, these institutions would indeed degenerate into "a hatchery of parrots."

The one aspect of science that creationists will allow is the predictions that they claim for their model. But when students of the human prehistoric record produce evidence that shows such things as the gradual transformation of nonhuman to human form and the very late appearance of agriculture in the spectrum of human existence, the creationists either ignore the facts or deny that they can be assessed by the verifiable techniques of current science. Of course, they have every right to do and believe as they wish, but they do not have the right to enforce the teaching of their religiously based convictions in the public schools under the name of science.

REFERENCES CITED

Alexander, Richard D. 1978. Evolution, creation and biology teaching. *The American Biology Teacher*, vol. 40, no. 2, pp. 91–107.

Alter, Robert. 1981. *The art of biblical narrative.* New York: Basic Books.

Boule, Marcellin and Vallois, H. V. 1957. *Fossil men.* New York: Dryden Press.

Brace, C. Loring. 1973. Sexual dimorphism in human evolution. *Yearbook of Physical Anthropology* 16: 31–49.

—. 1977. Occlusion to the anthropological eye. In *The biology of occlusal development*, ed. James A. McNamara, Jr., pp. 179–209. Monograph no. 7, Craniofacial Growth Series. Ann Arbor, Mich.: Center for Human Development.

—. 1978. Tooth reduction in the Orient. *Asian Perspectives*, vol. 19, no. 2, pp. 203–19.

—. 1979*a*. Biological parameters and Pleistocene hominid lifeways. In *Primate ecology and human origins: ecological influences on social organization*, eds. Irwin S. Bernstein and Euclid O. Smith, pp. 263–89. New York: Garland Press.

—. 1979*b*. Krapina 'classic' Neanderthals, and the evolution of the European face. *Journal of Human Evolution*, vol. 8, no. 5, pp. 527–50.

—. 1979*c*. Review of *Origins* and *People of the Lake* by Richard E. Leakey and Roger Lewin. *American Anthropologist*, vol. 81, no. 3, pp. 702–04.

—. 1979*d*. *The stages of human evolution.* 2nd ed. Englewood Cliffs, N.J.: Prentice-Hall.

—. 1980. Australian tooth-size clines and the death of a stereotype. *Current Anthropology*, vol. 21, no. 2, pp. 141–64.

—. 1981. Tales of the phylogenetic woods: the evolution and significance of evolutionary trees." *American Journal of Physical Anthropology*, vol. 56, no. 4, pp. 411–29.

Brace, C. Loring and Mahler, Paul E. 1971. Post-Pleistocene changes in the human dentition. *American Journal of Physical Anthropology,* vol. 34, no. 2, pp. 191–204.

Brace, C. Loring, Mahler, P. E., and Rosen, R. B. 1973. Tooth measurements and the rejection of the taxon *'Homo habilis'. Yearbook of Physical Anthropology* 16:50–68.

Brace, C. Loring and Montagu, Ashley. 1977. *Human evolution: an introduction to biological anthropology.* 2nd ed. New York: Macmillan.

Brace, C. Loring, Nelson, Harry, Korn, Noel, and Brace, Mary L. 1979. *Atlas of human evolution.* 2nd ed. New York: Holt, Rinehart and Winston.

Butzer, Karl W., Isaac, Glynn Ll., Butzer, Elizabeth, and Isaac, Barbara, eds. 1975. *After the Australopithecines: stratigraphy, ecology, and culture change in the middle Pleistocene.* The Hague: Mouton.

Clark, W. E. Le Gros. 1950. *History of the primates.* London: British Museum.

—. 1960. *The antecedents of man: an introduction to the evolution of the primates.* Chicago: Quadrangle Press.

Cooke, H. B. S. 1976. Suidae from Plio-Pleistocene strata in the Rudolf basin. In *Earliest man and the environments in the Lake Rudolph basin,* eds. Y. Coppens, F. C. Howell, G. Ll. Isaac, and R. E. F. Leakey, pp. 251–63. Chicago: Univ. of Chicago Press.

Curtis, Garniss H. 1975. Improvements in potassium-argon dating 1962–1975. *World Archaeology,* vol. 7, no. 2, pp. 198–209.

Curtis, G. H., Drake, T., Cerling, T., and Hampel J. 1975. Age of KBS tuff in Koobi Fora formation East Rudolf, Kenya. *Nature* 258:395–98.

Dubois, Eugene. 1894. *Pithecanthropus erectus, eine Menschenähnliche Uebergangsform aus Java.* Batavia: Landesdruckerei.

Eldredge, Niles and Gould, Stephen J. 1972. Punctuated equilibria: an alternative to phyletic gradualism. In *Models in paleobiology,* ed. T. J. M. Schopf, pp. 82–115. San Francisco: Freeman, Cooper and Co.

Fairservis, Walter A., Jr. 1975. *The roots of ancient India: the archaeology of early Indian civilization.* 2nd ed. Chicago: Univ. of Chicago Press.

Fisher, Frank R., ed. 1961. *Man living in the Arctic.* Washington, D. C.: National Academy of Sciences—National Research Council.

Fleagle, J. G., Simons, E. L., and Conroy, G. C. 1975. Ape limb bones from the Oligocene of Egypt. *Science* 189:135–37.

Gingerich, Philip D. 1976. Paleontology and phylogeny: patterns of evolution of the species level in early Tertiary mammals. *American Journal of Science,* vol. 276, no. 1, pp. 1–28.

Gish, Duane T. 1974. *Have you been brainwashed?* Seattle: Life Messengers.

—. 1979. *Evolution? The fossils say no!* 3rd ed. San Diego: Creation-Life Pubs.

Gleibermann, L. 1973. Blood pressure and dietary salt in human populations. *Ecology of Food and Nutrition* 2:143–56.

Gould, Stephen Jay and Eldredge, Niles. 1977. Punctuated equilibria: the tempo and mode of evolution reconsidered. *Paleobiology,* vol. 3, no. 2, pp. 115–51.

Greenfield, Leonard O. 1972. Sexual dimorphism in *Dryopithecus africanus. Primates* 13:395–410.

—. 1977. *Ramapithecus and early hominid origins.* Ph. D. dissertation. Univ. of Michigan, Ann Arbor.

Grosch, Daniel S. and Hopwood, Larry E. 1979. *Biological effects of radiation.* 2nd ed. New York: Academic Press.

Holloway, Ralph L. 1975. Early hominid endocasts: volumes, morphology and significance for hominid evolution. In *Primate functional morphology and evolution,* ed. R. H. Tuttle, pp. 391–415. The Hague: Mouton.

—. 1980. The O.H. 7 (Olduvai Gorge, Tanzania) hominid partial brain endocast revisited. *American Journal of Physical Anthropology,* vol. 53, no. 2, pp. 267–74.

—. 1981*a.* The Indonesian *Homo erectus* brain endocasts revisited. *American Journal of Physical Anthropology,* vol. 55, no. 4, pp. 503–21.

—. 1981*b.* Volumetric and assymetry determinations on recent hominid endocasts: Spy I and II, Djebel Irhoud I, and the Salé *Homo erectus* specimens, with some notes on Neanderthal brain size. *American Journal of Physical Anthropology,* vol. 55, no. 3, pp. 385–93.

Howells, William W. 1974. Neanderthals: names, hypotheses, and scientific method. *American Anthropologist,* vol. 76, no. 1, pp. 24–38.

—. 1975. Neanderthal man: facts and figures. In *Paleoanthropology: morphology and paleoecology,* ed. R. H. Tuttle, pp. 389–407. The Hague: Mouton.

—. 1980. *Homo erectus*—who, when and where: a survey. *Yearbook of Physical Anthropology* 23:1–23.

Isaac, Glynn, Ll. 1976. The activities of early African hominids: a review of archaeological evidence from the time span two and a half to one million years ago. In *Human origins,* eds. G. Ll. Isaac and E. R. McCown, pp. 483–514. Menlo Park, Calif.: W. A. Benjamin.

Isaac, Glynn Ll. and McCown, Elizabeth R., eds. 1976. *Human origins: Louis Leakey and the East African evidence.* Menlo Park, Calif.: W. A. Benjamin.

Jenkins, Farish A., Jr. and Fleagle, J. G. 1975. Knuckle-walking and functional anatomy of the wrists in living apes. In *Primate functional morphology and evolution,* ed. Russell H. Tuttle, pp. 213–27. The Hague: Mouton.

Johanson, Donald C. and Edey, Maitland. 1981. *Lucy: the beginning of humankind.* New York: Simon and Schuster.

Johanson, Donald C. and White, Tim D. 1979. A systematic assessment of early African hominids (Primates: Hominidae). *Science* 203:321–30.

Johanson, Donald C., White, Tim D., and Coppens, Yves. 1978. A new species of the genus *Australopithecus* (Primates: Hominidae) from the Pliocene of eastern Africa. *Kirtlandia: The Cleveland Museum of Natural History* 28:-1–14.

—. 1982. Dental remains from the Hadar formation, Ethiopia: 1974–1977 collections. *American Journal of Physical Anthropology.* 57:545–604.

Jolly, Clifford J., ed. 1978. *Early hominids of Africa.* New York: St. Martin's Press.

Khatri, A. P. 1963. A century of prehistoric research in India. *Asian Perspectives,* vol. 6, nos. 1–2, pp. 169–85.

Leakey, L. S. B., Tobias, P. V., and Napier, J. R. 1964. A new species of the genus *Homo* from Olduvai Gorge. *Nature* 202:7–9.

Leakey, M. D., Hay, R. L., Curtis, G. H., Drake, R. E., Jackes, M. K., and White, T. D. 1976. Fossil hominids from the Laetolil beds. *Nature* 262:-460–66.

Leakey, Richard E. F. 1973. Evidence for an advanced Plio-Pleistocene hominid from East Rudolf, Kenya. *Nature* 242:447–50.

Leakey, Richard E. F. and Lewin, Roger. 1977. *Origins: what new discoveries reveal about the emergence of our species and its possible future.* New York: Dutton.

Lee, Richard B. and DeVore, Irven, eds. 1968. *Man the hunter.* Chicago: Aldine.

Livingstone, Frank B. 1958. Anthropological implications of sickle cell gene distribution in West Africa. *American Anthropologist,* vol. 30, no. 3, pp. 533–62.

—. 1973. *Data on the abnormal hemoglobins and glucose-6-phosphate dehydrogenase deficiency in human populations. Constricutions in human biology to archaeology 1.* Technical Report No. 3. Ann Arbor: Univ. of Michigan Museum of Anthropology.

Lovejoy, C. Owen. 1974. The gait of Australopithecines. *Yearbook of Physical Anthropology* 17:147–61.

—. 1975. Biochemical perspectives on the lower limb of early hominids. In *Primate functional morphology and evolution,* ed. R. H. Tuttle, pp. 291–326. The Hague: Mouton.

—. 1978. A biomechanical review of the locomotor diversity of early hominids. In *Early hominids of Africa,* ed. Clifford J. Jolly, pp. 403–29. London: Duckworth.

Lovejoy, C. Owen, Heiple, Kingsbury G. and Burstein, Albert H. 1973. The gait of *Australopithecus. American Journal of Physical Anthropology,* vol. 38, no. 3, pp. 757–80.

Luckett, W. Patrick, ed. 1980. *Comparative biology and evolutionary relationships of tree shrews.* New York: Plenum Press.

Macfarlane, W. V. 1973. Functions of aboriginal nomads during summer. In *Human biology of aborigines in Cape York,* ed. R. L. Kirk, pp. 49–68. Australian Aboriginal Studies No. 44, Part II. Canberra: Australian Institute for Aboriginal Studies.

Mayr, Ernst. 1971. Evolution vs. special creation. *The American Biology Teacher,* vol. 33, no. 1, pp. 49–50.

Morris, Henry M. 1963. *The Twilight of Evolution.* Philadelphia: Presbyterian and Reformed Pub. Co.

—. 1974. Ed., *Scientific creationism* (public school edition). San Diego: Creation-Life Pubs.

—. 1975. *The troubled waters of evolution.* San Diego: Creation-Life, Pubs.

Pilbeam, David. 1968. The earliest hominids. *Nature* 219:1335–38.

—. 1980. Major trends in human evolution. In *Current argument on early man,* ed. Lars-König Königsson, pp. 261–85. New York: Pergamon Press.
Radinsky, Leonard. 1973. *Aegyptopithecus* endocasts: oldest record of a pongid brain. *American Journal of Physical Anthropology,* vol. 39, no. 2, pp. 239–48.
Reed, Charles A., ed. 1978. *Origins of agriculture.* The Hague: Mouton.
Ryan, Alan S. 1980. *Anterior dental microwear in hominid evolution: comparisons with human and nonhuman primates.* Ph. D. dissertation. Univ. of Michigan, Ann Arbor.
Sankalia, H. D. 1978. The early Paleolithic in India and Pakistan." In *Early Paleolithic in South and East Asia,* ed. Fumiko Ikawa-Smith, pp. 97–127. The Hague: Mouton.
Santa Luca, A. P. 1977. *A comparative study of the Ngandong fossil hominids.* Ph. D. dissertation. Harvard University: Cambridge, Massachusetts.
Simons, Elwyn L. 1967. The earliest apes. *Scientific American,* vol. 217, no. 6, pp. 28–35.
—. 1972. *Primate evolution: an introduction to man's place in nature.* New York: Macmillan.
Simpson, George Gaylord. 1953. *The major features of evolution.* New York: Columbia Univ. Press.
—. 1961. *Principles of animal taxonomy.* New York: Columbia Univ. Press.
So, Joseph K. 1980. Human biological adaptation to arctic and subarctic zones. *Annual Review of Anthropology* 9:63–82.
Steegmann, A. T. 1975. Human adaptation to cold. In *Physiological anthropology,* ed. A. Damon, pp. 130–66. Oxford: Oxford Univ. Press.
Thorne, Alan G. and Macumber, P. G. 1972. Discoveries of late Pleistocene man at Kow Swamp, Australia. *Nature* 238:316–19.
Thorne, Alan G. and Wolpoff, Milford H. 1981. Regional continuity in Australian Pleistocene hominid evolution. *American Journal of Physical Anthropology,* vol. 65, no. 3, pp. 337–49.
Tuttle, Russell H. 1967. Knuckle-walking and the evolution of hominoid hands. *American Journal of Physical Anthropology,* vol. 26, no. 2, pp. 171–206.
—. 1969. Knuckle-walking and the problem of human origins. *Science* 166:-953–61.
—. 1975. Ed., *Paleoanthropology, morphology and paleoecology.* The Hague: Mouton.
Urbach, Frederick, ed. 1969. *The biological effects of ultraviolet radiation (with emphasis on the skin).* New York: Pergamon.
Walker, Alan. 1980. 2,000,001 B.C.—early human life in Plio-Pleistocene Africa. Ermine Cowles Case Memorial Lecture, 4 November 1980, Univ. of Michigan: Ann Arbor.
Ward, Steven C., Johanson, D. C., and Coppens, Y. 1982. Subocclusal morphology and alveolar process relationships of hominid gnathic elements from the Hadar formation: 1974–1977 collections. *American Journal of Physical Anthropology* 57:605–30.

Weidenreich, Franz. 1940. Some problems dealing with ancient man. *American Anthropologist,* vol. 42, no. 3, pp. 375–83.
—. 1947. Facts and speculations concerning the origin of *Homo sapiens. American Anthropologist,* vol. 49, no. 2, pp. 187–203.
Weiner, J. S. 1954. Nose shape and climate. *American Journal of Physical Anthropology,* vol. 12, no. 4, pp. 1–4.
White, Tim D. 1977. New fossil hominids from Laetoli, Tanzania. *American Journal of Physical Anthropology,* vol. 46, no. 2, pp. 197–230.
—. 1980. Additional fossil hominids from Laetoli, Tanzania: 1976–1979 specimens. *American Journal of Physical Anthropology,* vol. 53, no. 4, pp. 487–504.
White, Tim D. and Harris, J. M. 1977. Suid Evolution and the correlation of African hominid localities. *Science* 198:13–21.
White, Tim D. and Johanson, Donald C. 1982. Pliocene hominid mandibles from the Hadar formation, Ethiopia: 1974–1977 collections. *American Journal of Physical Anthropology.* 57:501–44.
Wolpoff, Milford H. 1968. Climatic influence on the skeletal nasal aperture. *American Journal of Physical Anthropology,* vol. 29, no. 3, pp. 405–24.
—. 1980*a.* Cranial remains of middle Pleistocene European hominids. *Journal of Human Evolution,* vol. 9, no. 5, pp. 339–58.
—. 1980*b. Paleoanthropology.* New York: Knopf.
—. 1981. Cranial capacity estimates of Olduvai hominid 7. *American Journal of Physical Anthropology* 56: 297–304

14

The Evolution of Bible-science

BY ROBERT J. SCHADEWALD

Introduction

For nearly two thousand years, various groups of dogmatists have tried to force the universe to fit their interpretation of Scripture. They have judged and rejected evidence and explanations according to the standard of their own religious beliefs. On scriptural grounds, some have rejected (and some continue to reject) the sphericity of the earth, the Copernican system, and the evolution of life on earth. In the last two centuries, flat-earthers, geocentrists, and creationists have adopted a label for their dogmas: Bible-science. This term was embraced in nineteenth-century England, for example, by the flat-earth Bible-Science Defence Association and in twentieth-century America by the creationist Bible-Science Association.

Bible-scientists have waged war on conventional science, sometimes defending their beliefs with such potent arguments as the rack, rope, or stake. By the early nineteenth century, most Bible-scientists had resigned themselves to living on a spherical earth that orbits the sun. Then they were confronted with a triple threat. First, a mass of evidence had convinced most geologists that the earth was very old and that various forms of life had appeared (and sometimes disappeared) on it sequentially over a very long period of time. Secondly, geologists could find no evidence of a world-wide Deluge. And finally, in 1859, Charles Darwin presented a huge amount of evidence that life on earth had indeed evolved, and he proposed a theory that accounted for the proliferation of life on earth in terms of natural processes. Such ideas directly contradicted a literal interpretation of the first chapters of Genesis. The alarmed Bible-scientists launched an attack on conventional science that continues to this day.

"Scientific creationism" is the watchword of Bible-science today. Its ambitious aim is to reestablish Genesis as the ultimate authority in geology, biology, and cosmology. Since scientific creationism represents a continuation of a long tradition, it is difficult to assign a date to its genesis. Some trace its origin to the publication of *The Genesis Flood* by theologian John C. Whitcomb, Jr. and engineer Henry M. Morris in 1961. This book argues that the Noachian Deluge accounts for the geological evidence better than conventional geology. In it and subsequent books, the creationists have offered arguments to prove that the earth is only six thousand to ten thousand years old and that all forms of life were separately created.

Soon after *The Genesis Flood* was published, two of the major creationist organizations, the Bible-Science Association and the Creation Research Society, were formed. The third major group, the Institute for Creation Research, was organized in 1970 with Henry M. Morris as director. Individually and collectively, through books, pamphlets, and public lectures, these creationists took their message to their constituents.

In public, modern creationists march under the banner of science itself. In books and lectures intended for the skeptical public, they avoid mention of God and the Bible and make a great pother about science. Addressing believers, they tell a different story. They insist that modern geology and the theory of evolution are affronts to the Bible. In place of them they offer a complex pseudoscience they call "scientific creationism," of which "Flood Geology" is the central dogma.

Flood Geology is totally rejected by conventional geologists. As Whitcomb and Morris wrote, "Many thousands of trained geologists, most of them sincere and honest in their conviction of the correctness of their interpretation of the geologic data, present an almost unanimous verdict against the Biblical accounts of creation and the Flood. . . ." Nonetheless, they insisted that such professional opinions must be misguided, since "the instructed Christian knows that the evidences for full divine inspiration of Scripture are far weightier than the evidences for any fact of science. When confronted with the consistent Biblical testimony to a universal Flood, the believer must certainly accept it as unquestionably true" (1961, p. 118).

Why would all but a handful of fundamentalist geologists reject what is "unquestionably true"? Henry Morris suggested that the answer might be found in the Tower of Babel:

> Its top was a great temple shrine, emblazoned with zodiacal signs representing the host of heaven, Satan and his "principalities and powers, rulers of the darkness of this world" (Ephesians 6:12). These evil spirits there perhaps met with Nimrod and his priests, to plan their long-range strategy against God and His redemptive purposes for the post-diluvian world. This included especially the development of a non-theistic cosmology, one which could explain the origin and meaning of the universe and man without acknowledging the true God of creation. Denial of God's power and sovereignty in creation is of course foundational in the rejection of His authority in every other sphere.
>
> The solid evidence for the above sequence of events is admittedly tenuous. . . . If something like this really happened, early in post-diluvian history, then Satan himself is the originator of the concept of evolution. [Morris 1975, pp. 74–75]

"Solid evidence" or not, Morris's books repeatedly make the accusation that evolution is satanic. Though skeptics may question his story of the Revelation to Nimrod on Mount Babel, it is true that the idea of a changing, developing earth goes back to antiquity. It only became popular in Western thought, however, in the last few centuries.

The Old Earth

The roots of modern geology go back to Nicolaus Steno, a Danish cleric who in 1669 published a treatise on fossils that spelled out several principles of the formation of rock strata. Steno maintained that the fossils found in sedimentary rocks are the remains of animals killed in the Noachian Deluge. This opinion prevailed among geologists for another one hundred fifty years. In England, it was supported by Thomas

Burnet in *A Sacred Theory of the Earth* (1681), by John Woodward in *An Essay Toward a Natural Theory of the Earth* (1695), and by William Whiston, who suggested in *A New Theory of the Earth* (1696) that the Deluge was caused by a comet.

Meanwhile, some were speculating on the origin of animal species. In the mid-1700s, Comte Buffon, the eminent French naturalist, hit upon the idea of evolution by the variation of species and published it in his monumental *Natural History.* Buffon was promptly slapped down by the theologians at the Sorbonne University in Paris, who forced him to publish the following recantation:

> I declare that I had no intention to contradict the text of Scripture; that I believe most firmly all therein related about creation, both as to order of time and matter of fact. I abandon everything in my book respecting the formation of the earth, and generally all which may be contrary to the narrative of Moses. [White 1955, p. 215]

The seeds of the geological revolution were planted by Scottish geologist James Hutton in 1785, but it was Charles Lyell who made them grow. Lyell's *Principles of Geology* (1830) is considered by many the foundation of modern geology. By his time, few serious geologists accepted the idea of the universal Deluge. Indeed, Baron Cuvier, a French naturalist who was Lyell's principal adversary, suggested that the earth had endured *many* catastrophic floods, and his followers were called catastrophists. Cuvier did as much to destroy the concept of a single major and world-wide flood as did Lyell, who rejected catastrophism. Building on Hutton's ideas, Lyell argued that most of the earth's rocks were formed over a long period of time by natural processes observable to this day. Lyell's geology was called uniformitarianism.

The rocks of the earth's crust typically lie in layers that are or once were essentially horizontal. Clearly, these layers form a time sequence, with younger rocks above older. Most of the rocks are sedimentary, formed of particles that settled out of water or were carried by the winds. Such rock-forming processes are still going on, and their rates, though variable, can be estimated. Calculations based on these rates alone made it obvious that the rocks of our planet's surface required millions of years to form. (Much more reliable methods are available today to estimate the age of the rocks; see chapters in this volume by Abell and Brush.)

Furthermore, the fossilized remains of plants and animals found in the rocks had changed systematically with age. How could this have happened?

In 1843 Robert Chambers, a publisher with no formal scientific background, anonymously published *Vestiges of the Natural History of Creation.* Chambers tried to reconcile the Bible with uniformitarian geology, and the mechanism he suggested to account for the various forms of life on earth was essentially "evolution tempered by miracle" (White 1955, p. 65). The book was extremely popular, but the Bible-scientists attacked it as atheistic. Another attempt at limited compromise was Hugh Miller's *The Testimony of the Rocks* (1857). Noting the absence of evidence for a universal flood, Miller suggested that the Noachian Deluge was a local event in the Middle East.

It was into this atmosphere that Charles Darwin launched his theory of evolution by natural selection in 1859. A decidedly noncombative man who suffered from chronic illness, Darwin knew too well how the Bible-scientists had greeted books much less revolutionary than *The Origin of Species.* (Indeed, he had nurtured the theory for nearly twenty years before he published it.) His worst fears were justified; outraged theologians descended on him with fire and brimstone.

Few men have ever been so denounced, yet few men's works have ever won such rapid acclaim within the scientific community. Within a decade, most naturalists had accepted evolution. Outside the scientific community, it took longer. Still, by the turn of the century, most scholars and liberal theologians had made their peace with Darwin and with evolution (White 1955).

Among the conservatives, it was different. A deluge of antievolution books appeared during the century following publication of *The Origin of Species.* The most noteworthy output came from Seventh Day Adventist George McCready Price, who, beginning in 1913, turned out some twenty-five major antievolutionary works. Ironically, Darwinian evolution caused a regression among Bible-scientists. By about 1830 many conservatives were prepared to tolerate an ancient earth and a "day-age" theory of creation in which the Genesis "days" represent geologic ages. Now that the concept of an old earth was tied to a concept of changing life forms, Bible-scientists tried to jerk the rug out from under Darwin by throwing out the ancient earth along with evolution. As the evidence for evolution and for the great age of the earth continued to accumulate, a quirky conservatism evolved into a strident pseudoscience.

The Flood Theorists

The Great Deluge is the Grand Delusion of creationism, an albatross hung about their necks by the ancient Hebrew scribes who adapted an already-ancient Middle Eastern flood myth and recorded it in Genesis. The Deluge is for creationists the explanation for essentially all of the fossil-bearing rock now found on earth. On average, fossil-bearing rock strata cover the continents to a depth of about a mile. Among the difficulties that defy credible scientific treatment for the Deluge theory are:

1. The source of Flood waters
2. Explanation for the layering of different kinds of fossils in the rock strata
3. The sheer *quantity* of fossils
4. Structures obviously formed on land and now buried deep in the rock strata.

To account for the Deluge waters, some creationists propose that before the Flood the earth was surrounded by a canopy of water vapor. This basic idea was proposed in 1874 by Isaac Newton Vail, a Quaker Bible-scientist. Vail suggested that planets evolve through a ringed stage (like Saturn) to a canopied stage (like Jupiter) to a final earthlike condition. Modern creationists throw out Vail's evolution and keep his canopy. But none has ever shown how a canopy containing enough water to flood the earth could be stable or how the earth's creatures could survive the incredible atmospheric pressure it would cause.

The earth's fossils are known to occur in an almost completely exceptionless sequence or order.* That is, specific fossils are found only in rocks of a certain age. Trilobites, for instance, an ancient form of shelled sea animal, became extinct about 300 million years ago, so they are only found in older, generally deeper rocks. Other forms of shellfish are more modern and only appear in younger rocks, higher in the geologic col-

*The few exceptions are easily explained by geological disturbance, which can be detected without reference to the fossils; see chapters in this volume by Raup and Schafersman.

umn. Morris has argued the upwelling waters of the Flood sorted them by hydraulic drag (Whitcomb and Morris 1961; Morris 1974*b*). For objects of similar shape and density, however, the hydraulic drag force is proportional to cross-sectional area, while the gravitational force is proportional to volume. A rational creationist would therefore expect trilobites to be sorted according to size, with the large ones always deeper than the small ones. This is decidedly *not* the case, and one wonders how Morris, who has a Ph. D. in hydraulic engineering and a background in geology, could seriously advance an explanation based on hydraulic drag. In fact, if the great majority of the world's fossils were buried in the span of a single flood, as the creationists maintain, the mixing of life forms would have been absolutely phenomenal, and no orderly pattern could hope to emerge!

Beginning in 1961 with *The Genesis Flood*, numerous creationist books mention a huge rock body in Africa, the Karroo Formation, that contains the fossil remains of about 800 billion animals. Ironically, this formation alone totally refutes Flood geology. The animals of the Karroo range from the size of a small lizard to the size of a cow, the average being approximately fox-sized (Sloan 1980). Creationists claim they died in the Flood, so they must have been alive at its beginning. A simple calculation shows there are about *seven* animals in the Karroo formation alone for every acre of land on earth! This does not take into account the additional trillions fossilized in other formations, presumably also deposited during the Great Deluge.

Numerous features—wind-blown sand dunes, rain spatters, dinosaur nests, animal tracks, etc.—commonly found in the earth's rocks were clearly formed at the earth's surface. Yet they are found in rock strata creationists attribute to the Flood and with more such strata covering them.

Surely the present evidence against Flood geology is roughly equivalent to that which opposed the flat-earth theory when it flowered and flourished in the nineteenth century. Indeed, to understand Bible-science it helps to go back and examine its historic roots.

The Flat-earthers

The first great battle between Christianity and science involved the shape of the earth. The sphericity of the earth was well known in the Hellenistic culture in which Christianity developed. Long before, Aristotle had offered three proofs that the earth is a globe: (1) ships leaving port disappear over the horizon; (2) as one travels to the south, stars that are not visible in Greece appear above the southern horizon; and (3) during an eclipse, the earth's shadow on the moon is visibly curved. Unfortunately, this spherical model conflicts with the way the earth is depicted in the Bible.

The ancient Hebrews, like their older and more powerful neighbors, the Babylonians and the Egyptians, were flat-earthers. The Hebrew cosmology is never actually spelled out in the Bible but, even without knowledge of the Babylonian system upon which it is patterned, it can be read between the lines of the Old Testament. The Genesis creation story itself suggests the relative size and importance of the earth and the celestial bodies by specifying their order of creation. The earth was created on the first day, and it was "without form and void" (Genesis 1:2). On the second day a vault—the "firmament" of the King James Bible—was created to divide the waters, some being above and some below the vault. Not until the fourth day were the sun, moon, and stars created, and they were placed "in," not "above," the vault. The sizes of these bodies are not specified, but they had to be small, as Joshua later commanded the sun to stand still "in Gibeon" and the moon "in the Vale of Aijalon" (Joshua 10:12).

Other passages complete the picture. God "sits throned on the vaulted roof of earth, whose inhabitants are like grasshoppers" (Isaiah 40:21–22). He also "walks to and fro on the vault of heaven" (Job 22:14), which vault is "hard as a mirror of cast metal" (Job 37:18). The roof of the sky has "windows" (Genesis 7:12) that God can open to let the waters above fall to the surface as rain. The topography of the earth isn't specified, but Daniel "saw a tree of great height at the centre of the earth . . . reaching with its top to the sky and visible to the earth's farthest bounds" (Daniel 4:10–11). Such visibility would not be possible on a spherical earth, but might be expected if the earth were flat.

These passages, plus the slightly more explicit astronomical section in the influential but noncanonical Book of Enoch (Charles 1913), led the more literal-minded early Christians to reject the idea of a spherical earth as heresy. Thus flat-earthism has been associated with Christianity since the beginning. Many of the fathers of the Church were flat-earthers, including Lactantius, Tertullian, and Clement of Alexandria (White 1955, p. 92). Gradually they developed a "scientific" flat-earth system with which to oppose the Ptolemaic astronomy then becoming popular.

Claudius Ptolemy's *Almagest*, written in about 140 A.D., was the culmination of more than six centuries of Greek astronomy. It wasn't until about 550 A.D. that Cosmas Indicopleustes published the alternate (flat-earth) system in his book *Christian Topography*. Cosmas, an Egyptian monk, offered many of the same arguments used by flat-earthers today, but he came to a different conclusion about the shape of the earth. Using a barrage of quotations from Scripture and from the fathers of the Church, Cosmas tried to show that the earth is a rectangular plane with its east-west dimension twice the north-south dimension. Sunrise and sunset, he suggested, were caused by a huge mountain in the far north (Cosmas 550 A.D.).

Cosmas fought a losing battle, and the ancient idea of a flat earth quickly lost ground. The Ptolemaic system of astronomy, based on a spherical earth, worked reasonably well. By the twelfth century, flat-earthism was essentially a dead letter in the West.

While the Bible does not flatly state the shape of the earth, it repeatedly says in plain Hebrew that the earth is immovable (see, for instance, I Chronicles 16:30, Psalm 93:1, Psalm 96:10, and Psalm 104:5). Thus churchmen who found it easy to ignore its flat implications and adopt the geocentric but spherical system of Ptolemy were rudely shaken by Copernicus and Galileo. The Catholic church's reaction to Galileo is well known. It is less well known that most of the "reformers"—Luther, Calvin, Wesley—also rejected the Copernican system on scriptural grounds (White 1955, p. 126f). A few Protestant Bible-scientists have been fighting a rearguard action against heliocentricity ever since.

The modern flat-earth movement was launched in England, in 1849, with the publication of a sixteen-page pamphlet, *Zetetic Astronomy: A Description of Several Experiments which Prove that the Surface of the Sea Is a Perfect Plane and that the Earth Is Not a Globe!* by "Parallax." For the next thirty-five years, Parallax—his real name was Samuel Birley Rowbotham—toured England, attacking the spherical system in public lectures. His completely original system, still known to its adherents as "zetetic astronomy," is best described in his 430-page second edition of *Earth Not a Globe*, published in 1873 also under the pseudonym of Parallax (1873*a*).

The essence of zetetic astronomy is as follows: The known world is a vast circular plane, with the north pole at the center and a one-hundred-fifty-foot wall of ice at the "southern limit." The equator is a circle roughly halfway in-between. The sun, moon, and planets circle above the earth in the region of the equator at an altitude of perhaps six hundred miles. Their apparent rising and setting is an optical illusion caused by atmospheric refraction and the zetetic law of perspective. The latter law also explains why ships apparently vanish over the horizon when sailing out to sea. The moon is self-luminous, and it is occasionally eclipsed by an unseen dark body passing in front of it. The entire known universe is literally covered by the "firmament" (vault) so often referred to in the King James Bible.

Rowbotham and his followers were sowing fertile ground, for the flat-earth movement really caught on about 1860, just after Darwin's *The Origin of Species* was published. Conservative churchmen were excoriating geology, and those who believed geologists were misleading them had little problem believing the same of astronomers. As the most outspoken Bible-scientists of the day, the flat-earthers made "Zetetic Astronomy" a household word in Victorian England. Before the end of the nineteenth century, the movement spread to America and the rest of the English-speaking world. Few professional academics embraced it, though there were exceptions. Alexander McInnes of Glasgow University was a vehement flat-earther, as was Arthur V. White of the University of Toronto. Most of the flat-earthers who could boast "credentials" were clerics or engineers.

In America flat-earthism became a central doctrine of Wilbur Glenn Voliva's Christian Catholic Apostolic Church in Zion, Illinois. During the 1920s and 1930s, thousands of residents of Zion were at least nominally flat-earthers. In some families three generations learned the flat-earth doctrine in Zion parochial schools. From his one-hundred-thousand-watt radio station, Voliva used to thunder against "the Devil's triplets, Evolution, Higher Criticism* and Modern Astronomy." The popularity of flat-earthism has declined in America since Voliva's death in 1942, but the movement is alive and well and headquartered in Lancaster, California.

*Higher criticism is the study of the text of the Bible with the aim of elucidating its date, structure, sources, etc. It is anathema to most Bible-scientists.

Outcasts among Bible-scientists

Though flat-earthism is as well-supported scripturally and scientifically as creationism, the creationists plainly do not want to be associated with flat-earthers. In a public debate with Duane T. Gish, associate director of the Institute for Creation Research, paleontologist Michael Voorhies suggested that the Creation Research Society resembles the Flat Earth Society. According to a report of the debate published in the May 1979 issue of the creationist newsletter *Acts & Facts,* Gish replied "that not a single member of the Creation Research Society was a member of the Flat Earth Society and that Voorhies' linking of the two was nothing more than a smear." Gish's remarks brought a rejoinder in the September 1979 issue of *The Flat Earth News* from an outraged letter writer (identified only as "G. J. D.") who had read the *Acts & Facts* report. G. J. D. contested Gish's claim that no members of the Flat Earth Society belong to the Creation Research Society, concluding, "He doesn't know what he's talking about, as I belong to both, and I am writing to him to let him know that he is wrong." Ironically, Gish may have created a fact. To protest this attack on the flat-earthers, G. J. D. dropped out of the Creation Research Society.

Whether or not there are still flat-earthers in the Creation Research Society, scientific creationism closely resembles the flat-earth movement. In fact, scientific creationism, geocentrism, and flat-earthism are respectively the liberal, moderate, and conservative branches of the Bible-science tree. The intense hostility expressed by the scientific creationists towards the flat-earthers does not extend to modern geocentrists, who hover on the edge of respectability among creationists. Indeed, though the Bible is, from Genesis to Revelation, a flat-earth book, the geocentrists have combined forces with liberal creationists to cast the flat-earthers into outer darkness.

Despite their internecine warfare, Bible-scientists are in broad agreement on a number of scientific issues. They agree on the usefulness of the Bible as a scientific text, on the weakness of mere theories, and on the duplicity of conventional scientists.

The term "Bible-science" is meant in the most literal sense. To join the Creation Research Society, one must sign a statement of belief that begins: "The Bible is the written Word of God, and because we believe it to be inspired thruout, all of its assertions are historically and scientifically true in all of the original autographs. To the student of nature, this means that the account of origins in Genesis is a factual presentation of simple historical truths." Henry M. Morris puts it even more strongly:

> The real truth of the matter is that the Bible is indeed verbally inspired and literally true throughout. Whenever it deals with scientific or historical matters of fact, it means exactly what it says and is completely accurate. When figures of speech are used, their meaning is always evident in context, just as in other books. There is no scientific fallacy in the Bible at all. "Science" is *knowledge,* and the Bible is a book of true and factual knowledge throughout, on every subject with which it deals. The Bible *is* a book of science! [1974*a*, p. 229]

Samuel Birley Rowbotham, founder of the modern flat-earth movement, totally agreed with Morris: "To say that the Scriptures were not intended to teach science truthfully is, in substance, to declare that God himself has stated, and commissioned his prophets to teach, things which are utterly false" (Parallax 1873*a*, p. 357).

Similar sentiments were expressed by flat-earther David Wardlaw Scott, who wrote, "It [Scripture] never contradicts facts, and, to the true Christian student, it teaches more *real science* than all the schools and colleges in the world" (1901, p. 284). Elsewhere, Scott said of his book, "It may be that these pages will meet the need of some, who have not altogether been misled by unprovable fancies, and who will rejoice to find that the Biblical account of Creation is, after all, the only one which can be depended upon, and that Modern Astronomy, like its kindred theory of Evolution . . . is nothing but 'a mockery, a delusion, and a snare' " (1901, p. iii).

While theories are the backbone of conventional science, Scott's phrase "unprovable fancies" seems to epitomize what Bible-scientists think of them. They want nothing but the "facts." As Duane Gish once told an audience, "I have yet to find a scientific fact which contradicts the Bible, the Word of God. Now you and I are both aware of many scientific theories and opinions of scientific people that contradict the Scriptures. When we separate that which is merely opinion or theory or ideas from that which is established fact, there are no contradictions" (1978). Other creationists have expressed the same idea. In his preface to the creationist textbook *Biology: A Search for Order in Complexity,*

John N. Moore says that "true *science*" requires that the data "simply be presented *as it is*," and that "a philosophic viewpoint regarding *origins*" cannot be science (Moore and Slusher 1974).

Any flat-earther would agree. Indeed, in his lectures and writings, Samuel Birley Rowbotham repeatedly emphasized the importance of sticking to the facts. He called his system "zetetic astronomy" (zetetic from the Greek verb *zetetikos,* meaning to seek or inquire) because he sought only *facts* and left mere theories to the likes of Copernicus and Newton. Rowbotham devoted the entire first chapter of his *magnum opus* to praising facts at the expense of theories, concluding, "Let the practise of theorising be abandoned as one oppressive to the reasoning powers, fatal to the full development of truth, and, in every sense, inimical to the solid progress of sound philosophy" (Parallax 1873*a*, p. 8).

The fact is that Bible-scientists suspect theoretical scientists of duplicity. Scientific creationists rarely express their suspicions in plain English, but they strongly imply that much of modern science—radiometric dating, for instance—is a fraud. One prominent geocentrist, astronomer and computer scientist James N. Hanson, shows more candor. In a public lecture, he said of nongeocentric astronomers, "They lie a lot" (Hanson 1979). Charles K. Johnson, president of the Flat Earth Society, is absolutely vehement about scientific dishonesty. In the pages of *The Flat Earth News,* he regularly calls scientists "liars" and "demented dope fiends" and claims that the entire space program is a "carnie game."

Besides their agreement on scientific principles, many Bible-scientists share certain theological ideas. These are: (1) evolution (spherical theory) cannot be reconciled with the Bible, and accepting the former means rejecting the latter; (2) evolution (spherical theory) denies a personal God, leading to moral degeneration; and (3) it was for these reasons that Satan invented evolution (spherical theory).

Most Christians accept evolution as God's method of creation, a concept called "theistic evolution." Duane Gish has explicitly rejected that idea. "Not for a moment do I believe that the theory of evolution can be reconciled with the Bible. Theistic evolution is bankrupt both Biblically and scientifically. It's bad science and it's bad theology." And, "You really cannot believe the Bible and the theory of evolution both" (1978).

Rowbotham said the same of spherical theory: "Those Newtonian philosophers who still hold that the Sacred Volume is the Word of God, are . . . in a fearful dilemma. How can the two systems so directly opposite in character be reconciled?" (Parallax 1873*a*, p. 357). John Hampden put it more plainly: "No one can believe a single doctrine or dogma of modern astronomy, and accept Scriptures as divine revelation" (quoted by Rectangle 1899). Like all flat-earthers, Hampden also accepted the doctrine of creation in six solar days, and he extended the above opinion to the latter, writing, "If he can prove . . . that days do not mean days, then is the infidel fully justified in laughing to scorn every other phrase and every other statement, from the first verse to the last in the Bible" (Rectangle 1899).

Bible-scientists typically feel that orthodox science threatens the the doctrine of a personal God. As Duane Gish put it, "If they [conventional scientists] believe God exists, He's way out there somewhere, and has no real part in the origin of the universe" (1978). Henry Morris put it differently: "A great many people, particularly intellectuals, simply *prefer* an evolutionary theory of origins, because this device consciously or subconsciously relegates the Creator to a far-off, indefinite, or even illusory, role in the universe and in the lives of men who are in moral rebellion against him" (1963, pp. 92–93).

Albert Smith, once editor of the *Earth—Not a Globe—Review*, expressed the same insecurity as that felt by modern creationists: "On the astronomical hypothesis, the world is like an uncared-for orphan, or a desolate wanderer: God is removed too far from us to be of any practical use; and the idea of Heaven is so vague, that such a place, if it exist at all, may be anywhere or nowhere" (quoted in Rectangle 1899, p. 161).

Earlier, Henry M. Morris was quoted suggesting that Satan revealed evolution to Nimrod at the Tower of Babel. Elsewhere, he has repeatedly claimed that evolutionists are guided by the hand of Satan, whose concept of evolution is even older than Babel. "Behind both groups of evolutionists [theistic and nontheistic] one can discern the malignant influence of 'that old serpent, called the Devil, and Satan, which deceiveth the whole world' (Revelation 12:9). As we have seen, it must have been essentially the deception of evolution which prompted Satan himself to rebel against God, and it was essentially the same great lie with which he deceived Eve, and with which he has continued to 'deceive the whole world' " (1963, p. 93).

Morris's words would sound familiar to flat-earthers. "I believe the real source of Modern Astronomy to have been SATAN," wrote flat-earther David Wardlaw Scott (1901, p. 287). "From his first temptation of Eve in the Garden of Eden until now, his great object has been to throw discredit on the Truth of God." John Hampden agreed, calling the spherical theory "that Satanic device of a round and revolving globe, which sets Scripture, reason, and facts at defiance" (1886, p. 60).

In their imagined battle against Satan, creationists have adopted exactly the same tactics British flat-earthers used a century ago. These are essentially political tactics, aimed directly at the grassroots by way of the churches. Local activists sell books and pamphlets and write letters-to-the-editor. A corps of skilled lecturers crisscrosses the country speaking in churches, before religious organizations, or wherever they can get a hall and a crowd. In both the literature and the lectures the message is the same: the scientists are trying to destroy religion. Like England's Universal Zetetic Society, the Institute for Creation Research acts as publisher, lecture bureau, and communications clearing house for all these activities.

In the early 1970s members of the Institute for Creation Research began debating evolutionists whenever possible. In these debates, creationists take their text from flat-earther John Hampden, who once told a spherical critic, "We do not come before the public offering a kind of novelty and appealing for patronage. Nothing of the kind. You and the public are on the defence. We are your accusers. The novelty was introduced by you, and you are bound to justify your interference with what was perfect before" (1886, p. 64). Creationists expect their opponents to defend evolution while they throw rocks at it, and frequently opponents comply. The poor evolutionist, up against an experienced creationist debater, often looks like an unarmed man assaulting a fortress.

Many flat-earthers were also effective debaters. George Bernard Shaw described a public forum in which a flat-earther laid waste to the spherical opposition (Gardner 1957). Rowbotham was a tiger on the platform, and he was seldom bested. The good citizens of Leeds, England, once ran him out of town, being unable to make a more effective reply to his flat-earth arguments (Parallax 1873*b*). In Brockport, New York, in March 1887, two scientific gentlemen defended the sphericity of the earth against flat-earther M. C. Flanders on three consecutive nights.

When the great debate was over, five townsmen chosen to judge the matter issued a unanimous verdict. Their report, published in the Brockport *Democrat,* stated clearly and emphatically their opinion that the balance of the evidence pointed to a flat-earth (Hampden 1887).

Cash challenges are another way Bible-scientists taunt opponents. At the turn of the century, Koresh (Cyrus Reed Teed), a Chicago Bible-scientist, had a standing offer of $5,000 to anyone who could disprove his theory that the earth is hollow and we live on the inside of it. No one ever collected. In the 1920s and 1930s, Wilbur Glenn Voliva had a standing offer of $5,000 to anyone who could prove to him that the earth is not flat. No one ever collected. At the time of this writing (1982), creationist engineer R. G. Elmendorf offers $5,000 to anyone who can prove to him that evolution is possible (1976). Since Elmendorf is also something of a geocentrist, he offers $1,000 to anyone who can prove that the earth moves (1980).

Conclusion

In some respects, scientific creationists differ from most other Bible-scientists of the past century. Their doctrine has an emotional appeal not found in flat-earthism and geocentricity. Though scientifically hollow, it cannot be obviously falsified by something as simple as a shot into space. Most importantly, scientific creationists have significant political power, which they are eager to use. At this writing, Louisiana has a law on the books that would require that the scientific creationists' doctrines be taught in public schools whenever the theory of evolution is taught.* Significant agitation is going on in other states.

Thus the doctrines of Bible-science have evolved, but other aspects of the movement have gone full circle. While the rack, rope, and stake are fortunately obsolete, the Bible-scientists have once again given their doctrines the force of law.

*On November 22,1982, the Louisiana equal-time for creation science law was overturned on the grounds that the state legislature does not have the authority to determine school curricula.

REFERENCES CITED

Charles, R. H. 1913. *The Apochrypha and pseudepigrapha of the Old Testament,* vol. 2. Oxford: Clarendon Press.
Cosmas Indicopleustes. ca. 550 A.D. *Topographia Christiana.* Trans. J. W. McCrindle. London: Hakluyt Society (1897).
Elmendorf, R. G. 1976. $5,000 reward and a challenge to evolution. (Flyer dated 1 September 1976.)
—. 1980. $1,000 reward for scientific proof-positive that the earth moves. (Flyer dated 10 March 1980.)
Gardner, Martin. 1957. *Fads and fallacies in the name of science.* New York: Dover.
Gish, Duane T. 1978. Tape of a lecture to the Luthern Evangelistic Conference in Minneapolis, 23 January 1978.
Hampden, John. 1886. *The earth; Scripturally, rationally, and practically described. A geographical, philosophical, and educational review, nautical guide, and general student's manual,* #8. 11 December.
—. 1887. *The earth; Scripturally, rationally, and practically described. A geographical, philosophical, and educational review, nautical guide, and general student's manual,* #17. 1 November.
Hanson, James N. 1979. Tape of geocentric lecture delivered in Texas.
Moore, John N. and Slusher, Harold S. 1974. *Biology: a search for order in complexity.* Rev. ed. Grand Rapids, Mich.: Zondervan.
Morris, Henry M. 1963. *The twilight of evolution.* Grand Rapids, Mich.: Baker Book House.
—. 1974*a*. *Many infallible proofs.* San Diego: Creation-Life Pubs.
—. 1974*b*. *Scientific creationism.* San Diego: Creation-Life Pubs.
—. 1975. *The troubled waters of evolution.* San Diego: Creation-Life Pubs.
New English Bible. 1970. Oxford: Oxford Univ. Press.
Parallax (pseudonym of Samuel Birley Rowbotham). 1949. *Zetetic astronomy: a description of several experiments which prove that the surface of the sea is a perfect plane and that the earth is not a globe!* Birmingham, Eng.: W. Cornish.
—. 1873*a*. *Earth not a globe.* London: John B. Day.
—. 1873*b*. *The zetetic,* vol. 2, no. 2, pp. 39.
Rectangle (pseudonym of Thomas Winship). 1899. *Zetetic cosmogony; or conclusive evidence that the world is not a rotating, revolving globe, but a stationary plane circle,* 2nd ed. Durban, Natal (South Africa): T. L. Cullingworth.
Scott, David Wardlaw. 1901. *Terra firma: the earth not a planet.* London: Simpkin, Marshall and Co.
Sloan, Robert. 1980. Personal communication.
Whitcomb, John C. and Morris, Henry M. 1961. *The Genesis flood.* Nutley, N.J.: Presbyterian and Reformed Pub. Co.
White, Andrew D. 1955. *A history of the warfare of science with theology in Christendom,* vol. 1. New York: George Braziller. (First published in 1900.)

15

Is It Really Fair to Give Creationism Equal Time?

BY FREDERICK EDWORDS

Introduction

The idea of giving a minority position a fair hearing along with that of "the establishment" seems as American as apple pie. Some creationists have capitalized upon the widespread appeal of this openhanded approach in an effort to have their views given "equal time" in public schools whenever the "establishment" scenarios of biology, earth science, astronomy, social science, and even history, are taught. Posing as Galileos under persecution from "orthodox" science, they appear as champions of fairness in a fight against "dogmatic" evolutionists who want to "ban God from the public schools."

But is it really the scientific community that rejects fairness or is it the creationists who take an undemocratic position? A careful look at some of the complexities of the creation/evolution controversy will reveal where unfairness and dogmatism actually lie.

Fairness to All Christians

According to the equal-time concept, all significant alternate viewpoints should be given a fair hearing. Militant creationists, however, maintain that in the matter of origins there are only two such viewpoints: theirs and all others. Their view is expressed well in Section 4(a) of the 1981 Arkansas law that mandated "balanced treatment" for both creation and evolution. Since the Arkansas law is nearly identical to bills that have been pressed in so many states since the mid-1970s and to the 1981 Louisiana law as well, it is instructive to quote its definition of "creation-science":

> Creation-science includes the scientific evidences and related inferences that indicate: (1) Sudden creation of the universe, energy, and life from nothing; (2) The insufficiency of mutation and natural selection in bringing about development of all living kinds from a single organism; (3) Changes only within fixed limits of originally created kinds of plants and animals; (4) Separate ancestry for man and apes; (5) Explanation of the earth's geology by catastrophism, including the occurrence of a worldwide flood; and (6) A relatively recent inception of the earth and living kinds.

Many of the creationist public school textbooks, such as *Scientific Creationism, Origins: Two Models, Biology: A Search for Order in Complexity,* the *Science and Creation Series,* and others, give more details. In them we learn that "creationism" includes the belief that the entire universe is only about ten-thousand-years-old, that almost all fossils were laid down in the strata by a world-wide flood that sorted them in that order, that the Grand Canyon was carved out by this flood in less than a year, that limestone caves and oil and coal deposits were formed just as rapidly, that the survivors of the flood first appeared near the site of Mount Ararat, and that human languages had a miraculous origin near Babylon.

This all adds up to a rather specific and narrow brand of creationism. It is certainly based on Christian fundamentalism, but only a small number of fundamentalist Christians would qualify as "creationists" under this framework. It excludes, for example, any references to evolution as God's method of creation. Such a view, in the eyes of these militant creationists, is merely a brand of evolution and therefore doesn't fill the demand for "equal time for creationism." It seems, then, that "two-model" education may be excluding the view of millions of religious people—indeed, people who would call themselves creationists—who view origins differently from the minority that is pushing the legislation and the textbooks. That minority viewpoint is known in theological circles as Special Creationism. To understand just how far this exclusion of some creationist views goes, it will be necessary to survey a number of the ways different creationists interpret the Bible and science.

1. *The day/age theory.* This is the position that each "day" in the first creation story of the Bible actually represents a "long period" instead of an ordinary twenty-four-hour solar cycle. Day/age literalists would expect the rocks to reveal six separate and distinct creation epochs and the order of created life forms to be the same as that described in Genesis 1. Millions of Christians take this position, particularly the Jehovah's Witnesses.

The Jehovah's Witness Watchtower book *Did Man Get Here by Evolution or by Creation?* seeks to debunk evolution and to present its own view of creation. It says:

> The Bible gives no specific time period to the actual creating of the earth. Of the material universe, including the earth, the simple Bible statement is: 'In the beginning God created the heavens and the earth' (Genesis 1:1). This allows for thousands of millions of years that the material of the earth could have been in existence before being inhabited by living things. After this the Bible tells of six 'days' during which life appeared. But the Bible's use of the word 'day' here means a period of time and not a twenty-four-hour day. [Anonymous 1967, p. 97]

This position is further clarified in a discussion of radiometric dating methods.

> There are other dating methods too, but none in any way disprove the 6,000-year age of mankind given by the Bible. True, animal fossils are older, but the Bible, in its account of creation in Genesis, allows for that. It shows that animals were created thousands of years before man. [Anonymous 1967, p. 102]

So, although humans have been here only six thousand years, Jehovah's Witnesses believe animals and the rest of nature were created much earlier in a series of "long periods."

But the Special Creationists who are trying to legislate a requirement that their brand of creationism be taught whenever evolution is treated in science classrooms, reject the day/age theory as "unbiblical." Henry Morris, director of the Institute for Creation Research and editor of *Scientific Creationism* made this plain in a cover letter to the March 1980 issue of *Acts & Facts*.

> One of the greatest obstacles to the return of real creation teaching in our nation is the indifference of so many Christian people to the issue. They often justify this attitude on the basis of their assumption that people can believe in theistic evolution (or progressive creation) and still believe in the Bible. They feel that the evolutionary ages of geology can somehow be accommodated in Genesis, by means (usually) of the 'local flood' interpretation of the Noachian Deluge and the 'day/age' interpretation of God's week of creation.
>
> That honest and consistent Biblical exegesis excludes this interpretation is clearly demonstrated in . . . the enclosed March issue of *Acts & Facts.* I hope this study will encourage large numbers of sincere Christians everywhere to take a more forthright, Scripture-honoring stand on true creationism.

2. *The gap theory.* This view holds that there is a gap between the first and second verses in Genesis amounting to perhaps billions of years. In this gap is the first of two separate creations. This first creation involved all the extinct life forms found in the fossil record. According to the gap theory, the first creation was later destroyed and a second creation six thousand years ago brought about all life forms we see today (or that are *recently* extinct) in six solar days. Gap theorists expect the geological strata to reveal evidence of these two separate creation epochs. Many creationists support the gap theory, including the Worldwide Church of God, publishers of *The Plain Truth* magazine. Robert A. Ginskey, in his 1977 *Plain Truth* article, lays it on the line:

> The fact is, fundamentalists face a real problem in trying to squeeze dinosaurs into 6,000 years of earth history. The *facts* just don't allow it, even when Noah's Flood is invoked as an explanation. . . . Yet the Bible does supply the answer! It recognizes an inhabited earth—a prehuman world—*prior* to the creation week of Genesis 1. Genesis 1:1 speaks of the original creation: 'In the beginning God created the

> heaven and the earth.' But between the first two verses of Genesis, a time gap of unknown length exists. Surprising as it may seem, the Bible nowhere specifies the time of the original creation; perhaps it was billions of years ago. [pp. 30–31, 33, 41]

In this article Ginskey argues from scientific evidence to show why dinosaurs could not have lived recently and at the same time as humans (as Special Creationists claim). He notes the vast difference between the world of dinosaurs and that of human beings, describes how dinosaur fossils are not intermixed with those of modern mammals, presents some data from the various dating methods that show dinosaur fossils are very old, argues that the fossil record simply will not support the notion that the dinosaurs were all killed in Noah's flood, and quotes the Bible to demonstrate how wrong Special Creationists are. There is no doubt that this brand of creationism is just as much concerned with science and just as much concerned with religion as is Special Creationism.

Nonetheless, Henry Morris seems to show little tolerance for this position. In *Biblical Cosmology and Modern Science* he vigorously declares, "It is high time that Christians face the fact that the so-called geologic ages are essentially synonymous with the evolutionary theory of origins. That latter in turn is, at its ultimate roots, the anti-God conspiracy of Satan himself!" (1970, p. 71). Clearly, if Morris continues to have his way, the gap theory will never get equal time or be treated fairly as a creation model in the public schools.

3. *Progressive Creation.* This is the theory that there was a series of numerous separate creations, each interrupted by a gap in the fossil record. Progressive Creationists use the same fossil gap evidence and arguments that Special Creationists, like Morris, use—only Progressive Creationists use it more consistently. (Morris must posit a world-wide flood to stack up the strata. The Progressive Creationists can have a universal flood too, but because of their larger time frame they don't need it to do quite as much deceptive labor.) Progressive Creationists expect the fossil record to reveal no transitions, and they reject the notion of evolution as vehemently as do other creationists. Many religious people accept variations of Progressive Creation.

4. *Creation from chaos.* This is the view that God did not create the universe *ex nihilo* (from nothing) but rather formed it out of preexisting matter. Joseph Smith, the founder of the Mormon Church, took this

position in the King Follett sermon when he said, "Now the word create came from the word *baurau,* which does not mean to create out of nothing; it means to *organize;* the same as a man would organize materials and build a ship. Hence we infer that *God had materials to organize the world out of chaos*—choate matter, which is element, and in which dwells all the glory" (Swyhart 1976, p. 70). But since the Arkansas law and other creationist pieces of legislation have consistently defined creationism as meaning "sudden creation of the universe, energy, and life from nothing," it seems that many Mormons who would otherwise support creationism will be left out. Though not all Mormons are creationists, those who are, as well as those who are not, may find "two-model" education offensive.

Still other creationist interpretations of the Bible are logically plausible. For example, it is possible to suggest that after God first created the basic Genesis "kinds" many of them could have evolved greatly over time. There is nothing in the Bible to forbid this. Nor is there anything in the Bible to forbid the notion that humans and apes are related. Apes are not even mentioned in the Bible. Therefore apes and humans could be interpreted as one Genesis "kind" without sacrificing the notion of biblical inerrency. But Special Creationists will have none of this. In their view, Special Creationism is the only scientific, philosophical, and religious world view that represents "true" creationism. All other positions are merely different evolutionary schemes or compromises with evolutionary ideas. Special Creationists argue that such ideas *cannot* count as part of the creationists' "equal time," but, if taught, should be credited as arguments for evolution instead!

Fairness to Other Religions

Christianity isn't the only religion, nor is it the only religion that claims an active role for a divine power in creating organisms. Consider Hinduism, for example. Special Creationists are fond of pointing to the fact that the Hindu idea of reincarnation is a kind of "spiritual evolution"; they thus treat all Hindu ideas as evolutionary. Hindus themselves, however, don't necessarily agree with this interpretation of their beliefs. The Hare Krishnas, a popular Hindu cult in this country, are very critical of scientific materialism and evolution (Bhaktivedanta Swami Prabhupada 1979). Though their views differ from those of the Special Creationists, they use many of the same creationist arguments to support their theory of "production." Krishnas talk about probability, "living fossils," the supposed nonexistence of transitional forms in the fossil record, the lack of *conceivable* transitional forms, the necessity of design and intelligence in nature, and the inability of scientists to turn matter into consciousness. This is all familiar material that bears an amazing resemblance to the ideas of Morris as well as to those expressed in *The Plain Truth*.

In *Back to Godhead* magazine Jñana Dasa explains the Krishna notion of "production":

> This theory proposes that biological forms do not arise from the spontaneous self-organization of matter, but rather under the direction of a superior intelligence. Furthermore, it suggests that life and consciousness are not material phenomena, the results of physiochemical reactions. Instead, they result from a distinct, irreducible, nonphysical principle or entity, which is present within the material body during an individual's lifetime, and whose departure from the body leads to the change called death. [1979, p. 27]

This entity later enters other bodies in the reincarnation process, itself neither having been born nor ever dying.

Nonetheless, the Krishna acceptance of an old earth and a cyclical universe renders Krishna creationism a brand of evolution in the eyes of Special Creationists. In fact, Special Creationists hasten to point out that "true" biblical Christianity is absolutely unique among world religions in being the only one advocating a beginning—that is, a creation from nothing (Kofahl and Segraves 1975, p. 159–61). Although the creationists are wrong in asserting the uniqueness of Christianity with regard to this point, the use of this argument seems to lend credence to the Special Creationist notion that only one brand of creationism deserves to get equal time and should get it at the expense of all other world views.

In defending his "Resolution for Equitable Treatment of Both Creation and Evolution," a model school board resolution, Henry Morris wrote:

> There are really only two scientific models of origins—continuing evolution by natural processes or completed creation by supernatural processes. The latter need not be formulated in terms of Biblical references at all, and is not comparable to the various cosmogonic myths of different tribes and nations, all of which are merely special

> forms of evolutionism, rather than creationism, rejecting as they do the vital creationist concept of a personal transcendent Creator of all things in the beginning. [1975, p. iv]

This seems to mean that Special Creation deserves equal time not only because it is an alternate view to evolution, but because it is the *sole* alternate view. It should come as no surprise that many religious people, both non-Christian and Christian, both fundamentalist and liberal, find this position arrogant. This is why they oppose the major "equal time" legislative efforts, school board resolutions, and law suits that stem from this belief. They recognize the unfairness inherent in them.

Fairness to Other "Sciences"

The point most often belabored by Special Creationists is that they wish to teach an alternate *science,* not an alternate religion, in the public schools. They suggest that since, as the Arkansas law put it, "there are scientists who conclude that scientific data best support creation-science and because scientific evidences and inferences have been presented for creation-science," their view should be offered to balance evolution. But there are "scientists" who conclude that scientific data best support astrology too, and flat-earth science, and pyramid power! In fact, "scientific" evidences and inferences have been presented for a tremendous array of explanations that lack credibility. Generally speaking, it is only those explanations that have been tested by peers in the courts of the scientific community that are taught in public schools. But the Special Creationists aren't seeking peer review. They seek to force their theory into public schools even though it was rejected by the scientific community. But since they *are* willing to break the peer review rules, surely all other rule-breaking "sciences" must be given equal time, too. And if scientific standing is not to be the criterion for judging what should be taught in science classes, but public demand *is,* then the entire science curriculum should be drastically altered. Astrology doubtless has more adherents than does the theory that the universe is a mere six- to ten-thousand-years-old. Therefore, it would be strange indeed (and unfair) if creationism makes it into the public school science curriculum but astrology does not.

Besides those sciences that deal with the nature of the universe, there are sciences that deal with other subjects as well. One alternate science with religious overtones at least as great as those of creationism is Mary Baker Eddy's Christian Science theory of disease, which holds that illness is not caused by germs but rather by mental fixations or negative thoughts. Christian Scientists feel they can prove this and are probably religiously offended when the germ theory is presented in public schools as if it were the truth. And many *non*–Christian Scientists reject the germ theory, often basing their arguments on completely nonreligious considerations and evidences. Should this view get equal time whenever the school nurse comes in to lecture on hygiene or on how to prevent tooth decay?

And after covering biology and health classes, we should take a look at what is in store for geography teachers. Flat-earth science isn't much more outdated than creationism, and its supporters certainly understand its biblical base. Charles K. Johnson, president of the fifteen-hundred-member International Flat Earth Research Society, makes it very plain that the aim of the Bible is a "one world, flat-Earth society, for honesty and decency and that sort of thing." So Johnson should certainly get equal time on religious grounds. As far as science goes, he can offer many seemingly clear evidences to support his position. For example, anyone can see that water in a tub is flat. Therefore, if you expand on what you see right in front of you, you can only conclude that the whole earth has to be flat as well! The earth is a disc-shaped plane, Johnson argues, and there is even *experimental* evidence for this conclusion. In Columbus's famous 1492 test, three ships sailed the seas to the New World. Did Columbus fall off the earth? Not at all, which demonstrates conclusively that the earth cannot be a globe! Nonetheless, even the best of experiments must be repeatable, so Johnson's wife Marjory sailed to America from Australia and later swore in an affidavit that she never "hung by her feet in Australia," did not get on the ship upside down, and "did not sail straight up." She sailed directly across the ocean. Johnson considers this a very important proof. If it sounds absurd to the rest of us, it is probably because we have been raised on the globular hypothesis without ever getting the chance to hear the scientific evidences in favor of his view. "We consider this the world's most superstitious age," Johnson states. "We try to get people to use their minds logically" (Schadewald 1977, pp. 42–43; *Ashland Tidings* 1978).

Fairness in Other Classroom Subjects

Contrary to many popular assumptions, creationism does not concern itself only with science. The 1981 Arkansas law listed other courses of study where creationism might have a place alongside the evolutionary view.

> The subject of the origin of the universe, earth, life, and man is treated within many public school courses, such as biology, life science, anthropology, sociology, and often also in physics, chemistry, world history, philosophy, and social studies. [Section 7(a)]

The law therefore required that

> balanced treatment to these two models shall be given in classroom lectures taken as a whole for each course, in textbook materials taken as a whole for each course, in library materials taken as a whole for the sciences and taken as a whole for the humanities, and in other educational programs in public schools, to the extent that such lectures, textbooks, library materials, or educational programs deal in any way with the subject of the origin of man, life, the earth, or the universe. [Section 1]

As if in preparation for this law and similar school board resolutions, Creation-Life Publishers in 1976 came out with *Streams of Civilization*, a creationist *history* textbook designed for public school use. It "begins with creation, rather than cave man," says one advertisement. It offers "sound Christian teaching of history. The great men of the scriptures take their rightful places. Presents Noah and the flood as historical fact. Shows Jesus as more than just a man," says another (Institute for Creation Research 1976, p. 8; 1980, p. 8). One of the authors of this textbook, Mary Stanton, wrote in the March 1977 *Acts & Facts* that teleology, or God's guiding purpose, must be included in history studies. She therefore sees history and the Bible as "two of God's media for revelation." But she didn't stop there. As if to open as big a can of worms as possible, she pointed an accusing finger at historians who "continue to make Rome the first center of the Church" and who "give credit to Rome for establishing the solid foundation of Christianity and for spreading the Gospel during the first centuries after Pentecost." *Streams of Civilization* changes the emphasis to the Byzantines, a factor that will guarantee a counter equal-time demand from Roman Catholics.

But if we want to get into revisionist history, why limit it only to the disagreements among Christians, Protestant or Catholic? Why not give equal time to the Theosophists who teach that Jesus was instructed by Buddhist monks or to the Rosicrucians who hold that the greatest thinkers and spiritual leaders of all time possessed a secret knowledge passed down from ancient Egyptian priests through a secret society? And we cannot leave out Erich Von Däniken and his "ancient astronauts" theory of ancient history. Did beings from another world teach the Egyptians and Mayans how to build their pyramids? Or were these structures perhaps the products of people washed ashore after the sinking of Atlantis? Clearly, there is no lack of alternate ideas in history. There are both religious and secular brands of revisionist history, so if absolute fairness is the ideal, *all* of them should be included.

Fairness under the Constitution

In a surface reading the Arkansas law seems, in its wording, neither to mandate creationism nor to ban evolution. However, it actually does both. In Section 5 we read, "This Act does not require any instruction in the subject of origins, but simply requires instruction in both scientific models [of evolution-science and creation-science] if public schools choose to teach either." But the idea here expressed (essentially that the teaching of evolution is permissable only if the teaching of creation is equally included) is really two ideas in one. First, it is the idea that when evolution is taught, creation *must be taught as well.* Second, it is the idea that if creation is *not* taught, evolution *must be prohibited.* These two issues should be discussed separately. Let's begin with the second.

Prohibiting or banning evolution has a losing track record in the courts. Creationists have consistently tried to argue that evolutionary teaching or evolutionary museum exhibits are an affront to their religious beliefs and so should be removed, but this has been to no avail, at least as far as recent court decisions are concerned.

In the 1968 case of *Epperson* v. *Arkansas,* the U. S. Supreme Court held that no religious group had the right to blot out any public school teaching just because it was "deemed to conflict with a particular religious doctrine." In the 1978 case of *Crowley* v. *Smithsonian Institution*, the U. S. District Court declared that the creationists' free exercise of religion was "not actionably impaired merely because, should they visit the Smithsonian, they may be confronted with exhibits which are distasteful to their religion." And in the 1981 case of *Segraves* v. *California*, the judge ruled that California's guidelines on the exclusive teaching of evolution do not represent a burden on the religious rights of creationists. All the judge did in favor of creationists was to reaffirm that evolution could not be taught dogmatically or in a manner that attacked religion. But this had been established in the state nearly a decade before, and the state's guidelines do not break that rule.

This brings us to the first idea. Since banning evolution on religious grounds is an unconstitutional establishment of religion, it is all the more true that requiring a religious doctrine in the schools does the same thing! Therefore, requiring that creation be taught when evolution is presented amounts to such an establishment of religion. The *Epperson* decision addressed this point like it did the other. The Court declared:

> There is and can be no doubt that the First Amendment does not permit the State to require that teaching and learning must be tailored to the principles and prohibitions of any religious sect or dogma. . . . The State may not adopt programs or practices in its public schools or colleges which 'aid or oppose' any religion. . . . This prohibition is absolute. It forbids alike the preference of a religious doctrine or the prohibition of theory which is deemed antagonistic to a particular dogma.

This court ruling has helped insure that the schools are fair to all religions by seeing to it that they give favor to none.

It was on the basis of this court ruling, and others, that Judge William Overton, in *McLean* v. *Arkansas*, made his recent decision against the Arkansas creationism law. In his opinion he argued that "creation science" was indeed sectarian religion, declaring: "The evidence establishes that the definition of 'creation science' contained in [Section] 4(a) [of the Arkansas law] has as its unmentioned reference the first 11 chapters of the Book of Genesis" (1982, p. 17).

A counterargument creationists raise is that evolution is also a religion, the "religion of secular Humanism," and should therefore be dropped from the science curriculum as well. Evidence used to support this claim often consists of quotations from humanist materials, such as Humanist Manifesto II, which declares, "science affirms that the human species is an emergence from natural evolutionary forces." But the mere fact that humanists (much like Methodists, Presbyterians, Jews, Buddhists, and others) accept evolution does not make evolution Humanism. That some people rhapsodize about evolution and attempt to build an ethic based on it is really irrelevant. Humanists accept the germ theory of disease, too, but I doubt if creationists would want equal time given to the Christian Science theory for that reason. Humanists accept that the earth is round. Does this mean such an opinion is the "religion of secular Humanism" and must not be taught in the public schools? Perhaps this is why the U. S. District Court in *Crowley* v. *Smithsonian Institution* refused to accept this creationist argument. And, from a reverse viewpoint, many creationists, like humanists, oppose racism on religious grounds. Should racism be guaranteed equal time in schools because antiracism happens to correspond with many religious views?

It appears, then, that evolution is neither a burden on the religious rights of creationists nor a religion itself. On the other hand, creationism *is* religious and would therefore represent an unconstitutional establishment of religion if it were either mandated or if its criteria were used to ban evolution. The simple solution in cases where students are offended at evolution would seem to be discreet removal of the student from the class. Evolution is a purely secular study. But if it happens to offend someone's religion, just as saying the pledge of allegiance is an offense to Jehovah's Witnesses, that person has the right to leave. By no means, however, does that person have the right to prevent other students from learning about evolution or saying the pledge of allegiance. Such an act of prevention would only be justified if it could conclusively be shown that evolution or the pledge is a religion, and creationists have consistently failed to demonstrate this regarding the former.

In the process of defending their religious right to have creationism taught along with evolution, creationists seem to admit that creationism is a religion. Nonetheless, they also like to argue that it really isn't one. In the same Arkansas law that appealed to the free exercise of religion are these words: "Creation-science is an alternative scientific model of origins and can be presented from a strictly scientific standpoint without any religious doctrine just as evolution-science can. . . ." But even if this point is granted (which it was not in *McLean* v. *Arkansas*), there is no legal precedent for forcing any alternate science or pseudoscience into the school curriculum. The law does not meddle in the consensus of the scientific community and the details of curriculum. It would be making a dangerous and totalitarian move if it did. For example, when Lysenko promoted Lamarckian evolution in the USSR and banned the Darwinian variety, many leading Russian geneticists were sent to Siberia for not accepting this doctrine. Lysenkoism, however, could not solve crop breeding problems effectively (Gardner 1957), and Mendelian genetics proved necessary in the long run. The idea of legislating "truth" in science has repeatedly been tried and has shown itself to be a risky venture. The liberty of scientists to do their own research and publish their conclusions and the academic freedom of teachers to apply their professional expertise by teaching students what an independent science has discovered are important and practical rights that must always be gauranteed.

The Proper Place for Creationism

From the foregoing it can be seen that Special Creationists aren't really fighting for fairness but rather are fighting to have their religion taught in the public schools at taxpayer expense. After all, if they were really trying to be fair, they would seek to insure that *all* alternate religions and sciences achieved a portion of public school class time.

Of course, if all alternate religious views of origins were given an equal hearing in the science classroom, there would be no time left for the teaching of science itself. The result would be a course in comparative religion. This is fine, but it requires a specially trained teacher. Such teachers exist, and most scientists would have no objection to having courses on comparative religion included in the public school curriculums or to having Special Creation included in those curriculums.

As for alternative sciences, they are of interest to many students. Thus an elective course that presented astrology, pyramid power, "ancient astronauts," dowsing, flat-earth science, scientific creationism, and other "alternate" explanations, and allowed students to hear all sides of these issues, could be quite valuable. It would go well with instruction in critical thinking.

But it seems that every time courses on either comparative religion or critical thinking and the "fringes of science" are offered, the creationists object. They neither propose such courses themselves nor leave them alone if they are proposed by others. In light of this it is difficult to believe that most Special Creationists are genuinely fighting for fairness while most scientists are "dogmatic." The situation is actually the reverse.

The only thing scientists and civil libertarians are fighting for is the integrity of science itself. Just as English classes involve the teaching only of English, not French and Japanese, so science classes should involve the teaching of science only. If pseudoscience or religion is accorded "equal time," the impression is given that the scientific community is equally divided on the question of creation and evolution. To give this impression, which is false, is dishonest.

But scientists and civil libertarians do not wish to ban creationism from the public schools. They generally have no objection to comparative religion studies, where creationism would best be placed. Their objection is to *misplacing* creationism where it will only serve to confuse students about the real nature of the subject under study. They are trying to maintain the integrity of science, not ban a religion. So there is indeed a place for creationism in the public schools, a place where treatment of it will be truly fair. It just does not happen to be in the science classroom.

REFERENCES CITED

Anonymous. 1967. *Did man get here by evolution or by creation?* New York: Watchtower Bible and Tract Society of New York.

Arkansas, State of. 1981. Act 590 of 1981. 73rd General Assembly, Regular Session, March.

Ashland Tidings. 1978. 'Flat earth' fans say Columbus proved their claim. Ashland, Ore., 9 October.

Bhaktivedanta Swami Prabhupada, A. C. 1979. *Life comes from life.* Los Angeles: Bhaktivedanta Book Trust.

Dasa, Jñana. 1979. Evolution or production? *Back to Godhead,* vol. 14, no. 9, pp. 27–28, 30.
Gardner, Martin. 1957. *Fads and fallacies in the name of science.* New York: Dover.
Ginskey, Robert A. 1977. The Bible and the dinosaur world. *The Plain Truth* (May), pp. 30–33, 41.
Institute for Creation Research. 1976. Understandable history for your home! *Acts & Facts,* vol. 5, no. 9, p. 8.
—. 1980. Announcing a new arrival—*Streams of Civilization* volume II. *Acts & Facts* (March), p. 8.
Kofahl, Robert E. and Segraves, Kelley L. 1975. *The creation explanation.* Wheaton, Ill.: Harold Shaw Pubs.
McLean, Bill et al. vs. *Arkansas Board of Education et al.* 1982. Memorandum Opinion of Judge William Overton, 5 January 1982, United States District Court Eastern District of Arkansas.
Morris, Henry M. 1970. *Biblical cosmology and modern science.* Nutley, N.J.: Craig Press.
—. 1975. Resolution for equitable treatment of both creation and evolution. ICR Impact Series, no. 26. pp. i-iv.
—. 1980. Cover letter accompanying March 1980 *Acts & Facts.* San Diego: Institute for Creation Research.
Schadewald, Robert. 1977. The plane truth. *TWA Ambassador* (December), pp. 42–43.
Stanton, Mary. 1977. Can you recognize bias in history content? ICR Impact Series, no. 45, pp. 1–4.
Swyhart, Barbara Ann. 1976. *Narratives about cosmic and human origins: selected readings.* San Diego: San Diego City Schools.

Index